An Assessment of Atlantic Bluefin Tuna

Committee to Review Atlantic Bluefin Tuna

Ocean Studies Board

Commission on Geosciences, Environment, and Resources

National Research Council

NATIONAL ACADEMY PRESS
Washington, D.C.   1994

NOTICE: The project that is the subject of this report was approved by the Governing Board of the National Research Council, whose members are drawn from the councils of the National Academy of Sciences, the National Academy of Engineering, and the Institute of Medicine. The members of the committee responsible for the report were chosen for their special competencies and with regard for appropriate balance.

This report has been reviewed by a group other than the authors according to procedures approved by a Report Review Committee consisting of members of the National Academy of Sciences, the National Academy of Engineering, and the Institute of Medicine.

The National Academy of Sciences is a private, nonprofit, self-perpetuating society of distinguished scholars engaged in scientific and engineering research, dedicated to the furtherance of science and technology and to their use for the general welfare. Upon the authority of the charter granted to it by the Congress in 1863, the Academy has a mandate that requires it to advise the federal government on scientific and technical matters. Dr. Bruce Alberts is president of the National Academy of Sciences.

The National Academy of Engineering was established in 1964, under the charter of the National Academy of Sciences, as a parallel organization of outstanding engineers. It is autonomous in its administration and in the selection of its members, sharing with the National Academy of Sciences the responsibility for advising the federal government. The National Academy of Engineering also sponsors engineering programs aimed at meeting national needs, encourages education and research, and recognizes the superior achievements of engineers. Dr. Robert M. White is president of the National Academy of Engineering.

The Institute of Medicine was established in 1970 by the National Academy of Sciences to secure the services of eminent members of appropriate professions in the examination of policy matters pertaining to the health of the public. The Institute acts under the responsibility given to the National Academy of Sciences by its congressional charter to be an adviser to the federal government and, upon its own initiative, to identify issues of medical care, research, and education. Dr. Kenneth I. Shine is the president of the Institute of Medicine.

The National Research Council was organized by the National Academy of Sciences in 1916 to associate the broad community of science and technology with the Academy's purposes of furthering knowledge and advising the federal government. Functioning in accordance with general policies determined by the Academy, the Council has become the principal operating agency of both the National Academy of Sciences and the National Academy of Engineering in providing services to the government, the public, and the scientific and engineering communities. The Council is administered jointly by both Academies and the Institute of Medicine. Dr. Bruce Alberts and Dr. Robert M. White are chairman and vice chairman, respectively, of the National Research Council.

Support for this project was provided by the U.S. Department of Commerce, NOAA Contract No. 50-DGNC-3-00016.

Library of Congress Catalog Card Number 94-67949

International Standard Book Number 0-309-05181-9

Cover art by Ellen Hill-Godfrey. Ms. Hill-Godfrey received her master of fine arts degree from the University of North Carolina-Chapel Hill. Her paintings and prints have been exhibited in the Washington, DC, area and throughout the Mid-Atlantic and Southern regions. She has done illustrations for the University of Georgia Press and the University of North Carolina's *Endeavors* magazine. She lives in Germantown, MD, and teaches at The Barnesville School.

Copyright 1994 by the National Academy of Sciences. All rights reserved.

Printed in the United States of America

COMMITTEE TO REVIEW ATLANTIC BLUEFIN TUNA

John J. Magnuson, University of Wisconsin-Madison, *Chairman*
Barbara A. Block, Stanford University
Richard B. Deriso, Inter-American Tropical Tuna Commission
John R. Gold, Texas A&M University
William Stewart Grant, Wits University
Terrance J. Quinn II, University of Alaska
Saul B. Saila, University of Rhode Island
Lynda Shapiro, University of Oregon
E. Don Stevens, University of Guelph

Staff

Mary Hope Katsouros, *Director*
Robin Peuser, *Study Director*
Curtis Taylor, *Project Assistant*
Paulette Salmon, *Research Assistant*

OCEAN STUDIES BOARD

William Merrell, Texas A&M University, *Chairman*
Robert A. Berner, Yale University
Donald F. Boesch, University of Maryland
Kenneth Brink, Woods Hole Oceanographic Institution
Gerald Cann, Independent Consultant
Robert Cannon, Stanford University
Biliana Cicin-Sain, University of Delaware
William Curry, Woods Hole Oceanographic Institution
Rana Fine, University of Miami
John E. Flipse, Texas A&M University
Michael Freilich, Oregon State University
Gordon Greve, Amoco Production Company
Robert Knox, Scripps Institution of Oceanography
Arthur R. M. Nowell, University of Washington
Peter Rhines, University of Washington
Frank Richter, University of Chicago
Brian Rothschild, University of Maryland
Thomas Royer, University of Alaska
Lynda Shapiro, University of Oregon
Sharon Smith, University of Miami
Paul Stoffa, University of Texas

Staff

Mary Hope Katsouros, *Director*
Edward R. Urban, Jr., *Staff Officer*
Robin Peuser, *Research Associate*
Mary Pechacek, *Administrative Associate*
LaVoncyé Mallory, *Senior Secretary*
Curtis Taylor, *Office Assistant*
Paulette Salmon, *Research Assistant*

COMMISSION ON GEOSCIENCES, ENVIRONMENT, AND RESOURCES

M. Gordon Wolman, The Johns Hopkins University, *Chairman*
Patrick R. Atkins, Aluminum Company of America
Edith Brown-Weiss, Georgetown University Law Center
Peter S. Eagleson, Massachusetts Institute of Technology
Edward A. Frieman, Scripps Institution of Oceanography
W. Barclay Kamb, California Institute of Technology
Jack E. Oliver, Cornell University
Frank L. Parker, Vanderbilt/Clemson University
Raymond A. Price, Queen's University at Kingston
Thomas A. Schelling, University of Maryland
Larry L. Smarr, University of Illinois
Steven M. Stanley, The Johns Hopkins University
Victoria J. Tschinkel, Landers and Parsons
Warren Washington, National Center for Atmospheric Research

Staff

Stephen Rattien, *Executive Director*
Stephen D. Parker, *Associate Executive Director*
Morgan Gopnik, *Assistant Executive Director*
Jeanette Spoon, *Administrative Officer*
Sandi Fitzpatrick, *Administrative Associate*
Robin Allen, *Senior Project Assistant*

Preface

Issues of stock structure and international management of Atlantic bluefin tuna have been contentious for a number of years. In 1993, management issues became particularly antagonistic when several nongovernmental conservation organizations in the United States tried to have Atlantic bluefin tuna listed as an endangered species under the Convention on International Trade in Endangered Species Treaty and thus subject to severe trade restrictions.

In preparation for the 1994 annual management meeting of the International Commission for the Conservation of Atlantic Tunas (ICCAT), the U.S. Department of Commerce's National Oceanic and Atmospheric Administration (NOAA) asked the National Research Council (NRC) of the National Academies of Sciences and Engineering for independent advice on the scientific basis of management for Atlantic bluefin tuna. The NRC was asked to complete its work in six months so that this advice would be received in time to be useful for the ICCAT meeting.

The NRC's Ocean Studies Board formed the Committee to Review Atlantic Bluefin Tuna. The committee's charge was to:

1. Conduct a technical review and evaluation of the scientific basis of U.S. management of fisheries for Atlantic bluefin tuna, and address the following general questions:

    a. Are the current Standing Committees on Research and Statistics (SCRS) assessments of eastern and western Atlantic bluefin the most defensible interpretation of the available data?

b. Are uncertainties in the assessments dealt with adequately?

c. What is the status of the Atlantic bluefin tuna relative to the convention's goal of managing tuna to achieve maximum sustainable yield?

d. Does the available information support treating bluefin tuna as separate eastern and western management units (i.e., how much mixing is likely, and is it enough to invalidate two separate management units)?

2. Recommend research.

3. By August 30, 1994, produce for NOAA a report based on the committee's analysis of the technical information on the assessments of the status of eastern and western Atlantic bluefin tuna, including stock structure.

Acknowledgements to the many who helped the committee develop and complete this technical report and reanalysis follow this preface. However, I extend my personal thanks to an unusually strong and responsive committee and NRC staff, who made it work.

The full committee met twice to receive background presentations and draft its report. Several subgroup meetings also were held to work on the analyses and text of the report. The committee reviewed extensive background material,[1] including public comments that were received.

The committee intends this technical review to be of immediate use in management approaches and decisions by NOAA's National Marine Fisheries Service and ICCAT. Responding to the challenge implied by this scientific assessment will require a new view, attention to the best science, and a commitment to international cooperation. The committee dedicates this report to the giant bluefin tuna and to those who enjoy its majesty and value.

John J. Magnuson, *Chairman*
Committee to Review Atlantic Bluefin Tuna

---

[1] The report refers to several SCRS and ICCAT documents. For information on how to obtain any of these documents, contact NOAA's Office of International Affairs (301 713-2276) located at 1335 East-West Highway, Silver Spring, Maryland, 20910.

Acknowledgments

This study would not have been possible without the help and contributions of a number of individuals. The committee would especially like to thank Alejandro Anganuzzi, Richard Punsly, and Pat Tomlinson, who provided intensive technical support for the committee's analyses at the Inter-American Tropical Tuna Commission (IATTC) located in La Jolla, California. In addition, the committee thanks James Joseph of IATTC for allowing it to use the IATTC facilities and equipment for analyses and for hosting the final meeting in June. The committee extends its thanks to the other IATTC staff members who assisted in the June meeting, including Berta Juarez and Milton Lopez.

Staff of the Southeast Fisheries Science Center of the National Oceanic and Atmospheric Administration's (NOAA) National Marine Fisheries Service (NMFS) assisted in providing data sets for the stock assessment analyses. In addition, they provided the most recent International Commission for the Conservation of Atlantic Tunas (ICCAT) tag and release data for review. For the committee to have enough time for analyses, quick responses to requests for information and data sets were essential. The committee is appreciative of the prompt attention to its requests by NMFS's Southeast Fisheries Science Center staff and NMFS staff in Washington, D.C.

The committee also thanks Frank Hester and the East Coast Tuna Association for supplying the captains' logbook data that were analyzed in this report. In addition, the committee thanks Andre Punt for providing the software ADAPT used in the analyses for this report.

The committee expresses its gratitude to those who provided written comments and who at their own expense attended the open sessions of the committee meetings and provided oral comments. In particular, the committee thanks: Nelson Beideman, Gordon Broadhead, Paul Brouha, Philip G. Coates, Frank Cyganowski, John Mark Dean, Bill Degnan, Katsuma Hanafusa, Eric M. Hesse, Ken Hinman, John J. Hoey, Alexander Krause, Naozumi Miyabe, Thom Palchanes, Ellen M. Peel, Richard Ruais, Carl Safina, Stephen Sloan, Michael Sutton, Ziro Suzuki, Steve Weiner, and Peter Weiss.

Contents

LIST OF FIGURES AND TABLES		xv
SUMMARY		1
1	BACKGROUND	5
	ICCAT Regulations for Atlantic Bluefin Tuna, 6	
	Issues in Atlantic Bluefin Tuna Management, 7	
	Purpose of This Study, 8	
2	HISTORICAL EVIDENCE FOR STOCK STRUCTURE	9
	Introduction, 9	
	Concept of Stocks, 9	
	History of Atlantic Bluefin Tuna Stock Designations, 11	
	Genetic Studies, 13	
	Genetic Variation in Tunas and Scombroid Fish, 15	
	Genetic Variation in Bluefin Tuna, 16	
	Conclusion, 17	
	Recommendation, 17	
	Life History Parameters, 18	
	Geographic Locality of Spawning Grounds, 18	
	Timing of Spawning, 19	
	Age at Sexual Maturity, 20	
	Larval Biology, 21	
	Physiological Ecology of Bluefin Tuna Movement Patterns, 21	

Giant Bluefin Tuna in Winter, 28
Movements of Age 0 Fish, Small Fish, and Medium Fish, 28
Giant Bluefin Tuna in New England, 29
Conclusions, 29
Recommendations, 29
Climate, 30
  Climate and Evolutionary History, 30
  Conclusion, 31
  Fish Abundance and Climatic Changes, 31
  Conclusion, 35
  Recommendation, 35
Movement Based on Nongenetic Markers, 35
  Parasite Markers, 35
  Conclusion, 36
  Recommendation, 36
  Microconstituent Analysis, 36
  Conclusions, 36
  Recommendation, 37

3 TRANSATLANTIC MOVEMENT OF ATLANTIC
   BLUEFIN TUNA   39
   Introduction, 39
   Tag-Recapture Data, 39
   West to East, 39
   East to West, 41
   Reanalysis of Tagging Data, 42
   Methods, 42
   Results, 45
   Discussion, 76

4 FISH STOCK ASSESSMENT   79
   Growth, 79
   Standardization of Western Atlantic Bluefin Tuna Catch Rates
   (CPUEs), 80
     Introduction, 80
     Model Development, 80
     Analyses, 83
     Rod and Reel Indices for Small Fish, 83
     Rod and Reel Indices for Giant Bluefun Tuna, 84
     Captains' Logbook Data for Giant Bluefin Tuna, 86
     Comparison of Captains' Logbook Data and Rod and Reel
     Survey for Giant Bluefin Tuna, 89
     Trend Analysis, 91

Recommendations, 92
Quality Control, 92
Methods, 93
Population Assessment, 94
SCRS Base Case, 94
The Role of Indices, 95
Is the SCRS Base Case Reasonable?, 96
Are Uncertainties Adequately Incorporated into the SCRS Assessment?, 97
The Two-Area Mixed-Stock Case, 97
Discussion, 106
Conclusions—Standardization, 107
Recommendations—Standardization, 107
Conclusions—Population Assessment, 107
Recommendations—Population Assessment, 108

5   CONCLUSIONS AND RECOMMENDATIONS   109
General Recommendations, 109
Research Recommendations, 110

REFERENCES   113

APPENDIXES

A. Biographical Sketches of Committee Members   121
B. Bibliography   123
C. Genetic Variation in Other Tunas and Related Fish   127
D. Archival Tag Technology   135
E. Evidence for Mixing Based on Parasites   139
F. Microconstituent Analysis   147

List of Figures and Tables

FIGURES

2-1.	General distribution of bluefin tuna in the Atlantic Ocean.	12
2-2.	Landings of giant Atlantic bluefin tuna in Canadian waters.	26
2-3.	Map showing the localities of the Canadian catches from the Gulf of St. Lawrence and surrounding Atlantic Ocean regions.	27
2-4.	Barometric shifts and wind patterns driving the Russell Cycle.	32
3-1.	Proportion of tagged bluefin tuna in area 2 (eastern Atlantic Ocean) that were originally released in area 1 (western Atlantic Ocean) for a hypothetical population.	77
4-1.	Western Atlantic Ocean small bluefin tuna indices.	86
4-2.	Western Atlantic Ocean giant bluefin tuna indices.	89
4-3.	Observed and predicted values of indices for case 1.	103
4-4.	Comparison of the long-term trends in spawning stock biomass for cases 1, 2, and 7.	105
E-1.	The increase in size of *Nasicola* with host size in Atlantic bluefin tuna.	141
E-2.	Prevalence of parasites as a function of host age and locale of capture (A, *Nasicola*; B, *Elytrophora*).	143

TABLES

2-1.	Migration speed of Atlantic bluefin tuna calculated from tag-recovery data.	21
2-2.	Landings of giant Atlantic bluefin tuna from Canadian fisheries by province from 1960 to 1989.	24
3-1.	Synopsis of release and recapture (tagging experiments) of western Atlantic bluefin tuna.	40
3-2.	Synopsis of release and recapture (tagging experiments) of eastern Atlantic/Mediterranean bluefin tuna.	42
3-3.	Spanish tagging data for Atlantic bluefin tuna in the Cantabrian Sea (Bay of Biscay) from 1976 to 1991.	44
3-4.	Atlantic bluefin tuna release and recovery data from the United States tagging program in the western Atlantic Ocean. Tag returns are by year tagged and years out for all tag types.	46
3-5.	Atlantic bluefin tuna release and recovery data from the U.S. tagging program in the western Atlantic Ocean. Tag returns are by year tagged and years out for fish tagged with a single tag.	48
3-6.	Atlantic bluefin tuna release and recovery data from the U.S. tagging program in the western Atlantic Ocean. Tag returns are by year tagged and years out for fish tagged with two or more tags.	50
3-7.	Atlantic bluefin tuna tagged in the eastern Atlantic Ocean from 1976 to 1991. Month of tagging versus month of recapture, all types of tags.	52
3-8.	Atlantic bluefin tuna tagged in the western Atlantic Ocean from 1954 to 1990. Month of tagging versus month of recapture.	53
3-9.	Atlantic bluefin tuna release and recovery data from the U.S. tagging program in the western Atlantic Ocean. Tag returns are by year tagged and years out for single tags, double tags recovered with two tags, and double tags recovered with single tags.	54
3-10.	Atlantic bluefin tuna release and recovery data from the U.S. tagging program in the western Atlantic Ocean. Tag returns are by year tagged and quarters out for fish tagged with a single tag, tagged with two tags and recovered with two tags, and tagged with two tags and returned with one tag.	56
3-11.	Atlantic bluefin tuna tagged in the western Atlantic Ocean from 1971 to 1978. Estimates of annual instantaneous shedding rates using observed catches are shown in A, while B shows the observed catches increased by 20% for nonreporting.	58

LIST OF TABLES AND FIGURES   *xvii*

3-12. Atlantic bluefin tuna tagged with a single tag from 1971 to 1978. VPA analysis on the reported catches in number of fish, by time before recapture stratified by quarter-year intervals.   59

3-13. Atlantic bluefin tuna tagged with two or more tags from 1971 to 1978. VPA analysis on the reported catches, with two tags remaining, in number of fish by time before recapture stratified by quarter-year intervals.   60

3-14. Atlantic bluefin tuna tagged with a single tag from 1971 to 1978. VPA analysis on the reported catches in number of fish, increased by 20% to account for assumed non-reporting of tags, by time before recapture stratified by quarter-year intervals.   61

3-15. Atlantic bluefin tuna tagged with a two or more tags from 1971 to 1978. VPA analysis on the reported catches in number of fish with two tags remaining, increased by 20% to account for assumed non-reporting of tags, by time before recapture stratified by quarter-year intervals.   62

3-16. Atlantic bluefin tuna tagged in the western Atlantic Ocean with single, double, and multiple tags.   63

3-17. Atlantic bluefin tuna tagged in the western Atlantic Ocean with a single tag.   63

3-18. Atlantic bluefin tuna tagged with a single tag in the western Atlantic Ocean and recovered from the western Atlantic Ocean by year and by number of quarters out.   64

3-19. Atlantic bluefin tuna tagged with a single tag in the western Atlantic Ocean and recovered from the eastern Atlantic Ocean by year and by number of quarters out.   66

3-20. Atlantic bluefin tuna tagged in the eastern Atlantic Ocean and recovered in the eastern Atlantic Ocean versus time before recapture and a regression analysis of the natural logarithm of catch number against years out.   68

3-21. Atlantic bluefin tuna tagged in the western Atlantic Ocean from 1960 to 1981 with a single tag and recovered in the western Atlantic Ocean vs. years before recapture and a regression analysis of the natural logarithm of catch number against years out.   69

3-22. Atlantic bluefin tuna. VPA analysis on fish tagged in the eastern Atlantic Ocean from 1976 to 1991.   70

3-23. Atlantic bluefin tuna. VPA analysis on fish tagged in the western Atlantic Ocean from 1960 to 1981.   72

3-24. Estimates of transfer rates from east to west in the Atlantic Ocean, using tagging data.   74

xviii *LIST OF TABLES AND FIGURES*

3-25. Estimates of transfer rates from west to east in the Atlantic Ocean, using tagging data. 75

4-1. Estimates of small bluefin tuna CPUE from rod and reel and handline. 84
4-2. Relative abundance of small bluefin tuna from effort-weighted nonlinear least-squares regression. 85
4-3. Estimates of giant bluefin tuna CPUE from rod and reel and handline. 87
4-4. Relative abundance of giant bluefin tuna from effort-weighted nonlinear least-squares analysis. 88
4-5. Estimates of relative CPUE of giant bluefin tuna from captains' logbook data. 90
4-6. Trend analysis of abundance indices for western Atlantic bluefin tuna. 91
4-7. Trend analysis results for CPUE indices of Tables 1-5. 93
4-8. Spawning stock abundance and biomass ratios (1993/1988 and 1993/1975) for the western component of the Atlantic bluefin tuna. 99
4-9. Estimated exploitation rates by age in 1992 for the different cases. 100
4-10. Instantaneous fishing mortality rates as estimated for the different cases considered in the VPA. 101
4-11. Contribution of each index and total weighted sum of squared residuals for each of the cases considered in the VPA. 102

E-1. Prevalence of *Nasicola* sp. parasites in bluefin tuna related to age of the host and to locale. 142
E-2. Prevalence of *Elytrophora* sp. parasites in bluefin tuna related to age of the host and to the locale of host capture. 144
E-3. Prevalence of *Nasicola* sp., *Elytrophora* sp., and both parasites in the same host for western Atlantic bluefin tuna age 2 to 6. 144

F-1. Estimates of mixing between the two stocks based on discriminant function analysis using jackknife probabilities of group membership of adult bluefin tuna sampled from a variety of locations and at different times. 148

Summary

Atlantic bluefin tuna (*Thunnus thynnus*) are remarkable fish of considerable value, bringing a high selling price on the raw seafood market in Japan. Bluefin tuna are among the largest bony fishes in the ocean reaching lengths of over 10 feet (3.05 meters), weights of over 1,200 pounds (544 kilograms), and ages of over 30 years. These fish are prized catches of both commercial and recreational fishermen worldwide. The historic popularity of fishing for bluefin tuna and their increasing market value have contributed to the calamitous exploitation of this species, especially in the North Atlantic Ocean.

In response to concerns about the declining abundance of bluefin tuna in the North Atlantic Ocean during the mid-1960s, and in recognition of the need for coordinated international management of highly migratory fish species in the Atlantic Ocean, the International Convention for the Conservation of Atlantic Tunas was signed in 1966. The convention is implemented by an international body called the International Commission for the Conservation of Atlantic Tunas (ICCAT), presently consisting of 22 member nations, including the United States, Canada, Japan, Spain, and France. ICCAT is responsible for providing internationally coordinated research on the condition of Atlantic tunas and other large, highly migratory species (e.g., swordfish) and their environment, as well as for developing regulatory harvest proposals for consideration by member nations. The two most contentious issues concerning the management of Atlantic bluefin tuna are the definitions and sizes of management units and the indices of abundances now used in stock assessments.

In 1981, ICCAT adopted the premise of a two-stock structure for Atlantic bluefin tuna, one in the eastern and the other in the western Atlantic Ocean.

Implementation of this assumption began in 1982 for member nations. Since 1991, international management of the two stocks has been separated to the degree that each stock is alternately reviewed every second year by the commission. Because of the perceived decline in abundance of western Atlantic bluefin tuna, the two stocks have been subject to different regulations, the most striking difference being the absence of a quota for the eastern fishery and the imposition of a strict harvest quota, allowing catches only for scientific monitoring, for the western fishery.

The United States is represented in ICCAT by three commissioners. An official from the U.S. Department of Commerce's National Oceanic and Atmospheric Administration (NOAA) serves as one of the U.S. commissioners; of the other two, one is required to have experience with commercial fishing and the second with recreational fishing. In preparation for the November 1994 ICCAT meeting, Douglas K. Hall, Assistant Secretary for Oceans and Atmosphere in the Department of Commerce, asked the National Research Council's Ocean Studies Board to conduct a peer reviewed study within six months to enable NOAA to use the results for the ICCAT meeting.

Accordingly, the Ocean Studies Board established the Committee to Review Atlantic Bluefin Tuna to review and evaluate the scientific basis of U.S. management of fisheries for Atlantic bluefin tuna and to recommend research to resolve remaining stock structure issues. The report focuses primarily on the scientific basis for assumptions about stock structure and for indices of abundances used in the stock assessments for western Atlantic bluefin tuna. The issue of stock structure of Atlantic bluefin tuna is discussed in Chapter 2, and movement of Atlantic bluefin tuna is discussed in Chapter 3. Information on the indices of abundances and the results of sensitivity analyses conducted by the committee using data sets obtained from NOAA's National Marine Fisheries Service (NMFS) and the industry are presented and discussed in Chapter 4. In addition, the committee makes specific recommendations to address the detailed problems that are identified and discussed in each section of Chapters 2 through 4.

GENERAL RECOMMENDATIONS

In response to its charge and based on the analyses presented in Chapters 1 through 4, the committee presents in Chapter 5 the following major conclusions and recommendations for improving the scientific basis of management of Atlantic bluefin tuna:

1. Available biological evidence of stock structure although sparse is consistent with a single stock hypothesis for bluefin tuna in the North Atlantic Ocean, with at least two spawning areas. Furthermore, the committee's reevaluation of tagging results confirms that movement of bluefin tuna between the western and eastern Atlantic Ocean is sufficient to alter the previous ICCAT

Standing Committee on Research and Statistics (SCRS) stock assessments. **The committee recommends that NOAA/NMFS conduct new scientific assessments explicitly to include mixing of Atlantic bluefin tuna between eastern and western fishing grounds.**

2. In response to the first question posed to the committee by NOAA, the committee concludes that recent ICCAT SCRS assessments of abundance of eastern and western Atlantic bluefin tuna do not provide the most defensible interpretations of available scientific data. The committee's reanalyses show that there is no evidence that abundance of western Atlantic bluefin tuna has changed significantly between 1988 and 1992. **The committee recommends that NOAA/NMFS use alternative methods of data management, data analyses, and peer review for estimating abundance indices, movement rates, and mixed population assessments (as discussed in Chapters 3 and 4 of this report).**

3. The committee notes that the ICCAT SCRS uses a variety of uncertainty analyses. **The committee recommends that NOAA/NMFS and ICCAT SCRS act to include transatlantic movement of fish and adaptive management techniques in future uncertainty analyses.**

4. The committee cannot determine the maximum sustainable yield (MSY) for Atlantic bluefin tuna under a one-stock hypothesis with two spawning grounds. Available biological information on stock structure, mixing on the spawning and fishing grounds, spawning site fidelity, and spawner/recruit relationships is too sparse. We do know that the present abundance of bluefin tuna in the western Atlantic Ocean is lower than that in the early 1970s, although the committee did not analyze similar data for the bluefin tuna in the eastern Atlantic Ocean. We also know that the present abundance and fishing mortality are much higher in the eastern Atlantic Ocean than in the west, and that some physical mixing occurs between the fishing grounds in the eastern and western Atlantic Ocean. **The committee recommends that NOAA/NMFS reevaluate MSY for Atlantic bluefin tuna.**

RESEARCH RECOMMENDATIONS

The committee notes that research on the biology of Atlantic bluefin tuna is not continuing at an intensity necessary to answer major biological questions pertaining to the management of the fisheries. Therefore, the committee recommends that NOAA/NMFS carry out the research described below using the best available science and techniques within and outside NOAA. For example, research supported by other U.S. government agencies, including the National Science Foundation, the Department of Energy, and the Office of Naval Research, could contribute to the goals of the studies funded by NOAA. Finally, the committee urges NOAA/NMFS to work cooperatively with ICCAT to implement these research recommendations. The following recommendations are not listed in order of importance or priority.

1. Tagging data show that there is movement of bluefin tuna between the eastern and western Atlantic fishing grounds, but the degree of gene flow between spawning areas is not known. Such knowledge is essential in defining population genetic structure and useful for refining stock assessments. **The committee recommends that the one-stock hypothesis be tested rigorously, using the most appropriate technologies capable of detecting contemporary population genetic structure.**

2. Estimates of spawning fidelity to a particular area are essential for stock assessments. **The committee recommends that microconstituent analysis and archival tags be used to provide information on spawning fidelity.**

3. Stock assessments can be refined by better estimates of life history characteristics such as spawning biomass, larval abundance, sex ratio, age at maturity, fecundity, and recruitment. **The committee recommends that spawning biomass, sex ratio, age at maturity, and fecundity in the spawning grounds be estimated and that larval performance, as affected by environmental conditions, be studied.**

4. The committee recognizes that knowledge of movement patterns is essential for estimating abundance and distribution and that movement rates and patterns may change over time. **The committee recommends that a tagging program be undertaken, with an appropriate combination of conventional, PIT, acoustic, and archival tags to provide improved estimates of the magnitude and patterns of movement. This program should be designed to answer scientific questions pertinent to stock assessment and should be coordinated among all nations involved in the bluefin tuna fishery.**

5. Estimates of abundance are confounded by the interaction between fishing and changes in distribution caused by interdecadal climatic and oceanic variability. **The committee recommends a synthesizing analysis of existing data on distributions of bluefin tuna in relation to spatial and temporal dynamics of major oceanographic features.**

6. The committee notes that a greater use of peer review would have improved the quality of some of the research reviewed during the preparation of this report. **The committee recommends that review of all research proposals and resulting manuscripts include a process of external peer review.**

The committee believes that the analyses and results in this report present a challenge to government and to conservation and industrial organizations for the conservation and management of bluefin tuna in the Atlantic Ocean. Responding to this challenge will require a new view, better science, and a commitment to international cooperation. The committee hopes that this report will serve as a catalyst for obtaining better scientific information to improve the status of Atlantic bluefin tuna.

1

Background

Atlantic bluefin tuna (*Thunnus thynnus*) are prized catches of both recreational and commercial fishermen. These fish are migratory and are known to traverse the Atlantic Ocean in a few months. Bluefin tuna are among the largest bony fish in the ocean, reaching over 10 feet (3.05 meters) in length and over 1,200 pounds (544 kilograms) in weight. Their lifespans can exceed 30 years, making them long lived among fish species. The popularity of bluefin tuna, as demonstrated by the historic international fishery, has contributed to the significant exploitation of this species, especially in the North Atlantic Ocean. Furthermore, the high selling price of high-quality bluefin tuna on the raw seafood market in Japan has provided a financial incentive for the expenditure of great effort to pursue and catch these fish.

The need for coordinated international management of highly migratory fish species in the Atlantic Ocean was recognized in the mid-1960's, leading to the International Convention for the Conservation of Atlantic Tunas, signed on May 14, 1966. The convention is implemented by an international body called the International Commission for the Conservation of Atlantic Tunas (ICCAT), presently consisting of 22 member nations, including the United States, Canada, Japan, Spain, and France. The headquarters for ICCAT is in Madrid, Spain. ICCAT is responsible for providing internationally coordinated research on the condition of the Atlantic tunas and related species (e.g., swordfish) and their environment, as well as for the development of regulatory harvest proposals for consideration by the member nations. The objective of the regulatory proposals is to conserve and manage tuna and related species throughout their ranges in a manner that achieves the maximum sustainable catch.

The U.S. law implementing the convention is the Atlantic Tunas Convention Act of 1975 (ATCA). ATCA stipulates that the United States shall be represented in ICCAT by not more than three commissioners who are appointed by the President and who can serve no more than two three-year terms. One of the three U.S. commissioners can be a salaried government employee. To date, an official from the U.S. Department of Commerce's National Oceanic and Atmospheric Administration (NOAA) always has served as one of the U.S. commissioners. Of the other two, who are not employed by the government, one must be knowledgeable and experienced with regard to commercial fishing in the Atlantic Ocean, Gulf of Mexico, or Caribbean Sea, whereas the other is required to be knowledgeable and experienced with regard to recreational fishing in one of these regions.

The U.S. commissioners are assisted by an advisory committee consisting of between five and 20 individuals who are selected from the various groups concerned with fisheries that are governed by the convention. The advisory committee has the opportunity to offer comments on all proposed programs of investigation, reports, recommendations, and regulations of the commission.

ICCAT has four components: (1) the commission (composed of not more than three delegates from any member nation); (2) the council (an elected body with a chairman, vice-chairman, and representatives from four to eight member nations that performs functions assigned to it by the convention or commission); (3) the executive secretary (responsible for commission finances, coordinating ICCAT programs, preparing the collection and analysis of data to accomplish the purposes of the convention, and preparing reports for approval by the commission); and (4) subject area panels (established by the commission and responsible for reviewing the species under their purview, collecting scientific and other information, proposing recommendations for joint actions, and recommending studies by member nations). Standing Committees on Research and Statistics (SCRS) have been established by the commission.

The commission is responsible for formulating regulatory proposals, which are approved by ICCAT and submitted to member governments for approval. If there are no objections from any concerned contracting government within approximately six months, each party to the convention is then responsible for implementing and enforcing the regulations recommended by ICCAT.

ICCAT REGULATIONS FOR ATLANTIC BLUEFIN TUNA

In 1974, ICCAT recommended the first regulatory measures for Atlantic bluefin tuna. These measures included a minimum size limit of 6.4 kg and a limit on fishing mortality to the levels of 1974. In 1981, ICCAT adopted the premise of a two-stock structure for Atlantic bluefin tuna, one in the eastern and the other in the western Atlantic Ocean. Also in 1981 the capture of bluefin tuna in the western Atlantic Ocean was prohibited, except for a catch quota established specifically for continuing scientific monitoring of the western stock. This scientific catch was allocated to ICCAT-participating nations that had active fisheries (including the United States, Canada, and Japan). Brazil and Cuba were exempted from the regulations because their total catches were small (less than 50 metric tons).

For the 1983 fishing season in the western North Atlantic Ocean, ICCAT established a total allowable catch of 2,660 metric tons. ICCAT also limited the catch of smaller fish, less than 120 cm in length, to no more than 15% in weight of the catch limit for the western stock. In addition, spawning areas such as the Gulf of Mexico were protected by prohibiting any directed fishery there.

In each year from 1983 to 1991, ICCAT approved a one-year extension of the existing management measures for both stocks of bluefin tuna. During the 1991 ICCAT meeting, new regulations were recommended to reduce the western stock's scientific monitoring quota by 10% in 1992 and again in 1993, with the possibility of additional reductions of up to 25% based on future SCRS analyses. Other measures were adopted by ICCAT at the 1991 meeting, including limiting catches of small fish, penalties for exceeding quotas, promoting tag and release efforts, and a bluefin tuna documentation and reporting program phasing in. Exporters of bluefin tuna will be required to provide documents that identify the location and nation of the vessel that caught the fish. To date, ICCAT has not recommended a quota for the eastern stock. However, measures for catch size limits and protection of spawning areas have been implemented for the eastern stock.

ISSUES IN ATLANTIC BLUEFIN TUNA MANAGEMENT

ICCAT management for Atlantic bluefin tuna has evolved according to the premise of a two stock structure that can be managed independently. Implementation of this assumption began in member nations in 1982. Since 1991, each stock has been alternately reviewed every second year by the commission. Because of the perceived decline in abundance of western Atlantic bluefin tuna, the two stocks have been subject to different regulations, the most striking difference being the lack of a quota for the eastern stock fishery and the imposition of a strict scientific quota only for the western stock fishery. The two most contentious issues concerning the management of Atlantic bluefin tuna are the definitions and sizes of management units and the indices of abundances that are now used to calculate stock assessments. Opposition to managing the western and eastern Atlantic bluefin tuna stocks as separate units has arisen primarily from lack of definitive scientific evidence for genetically-discrete populations, and alternative population structures have been suggested. The implications for management are significant - if there are more or fewer populations, the present ICCAT management, including the regulations, would have to be modified and new regulations agreed to by all member nations.

PURPOSE OF THIS STUDY

The 1994 ICCAT meeting is scheduled for November in Madrid, Spain. Most management efforts will focus on the eastern stock, because this is its year for assessment. The western stock will be assessed at the 1995 ICCAT meeting. However, new measures could be recommended for the western stock in 1994 and for the eastern stock in 1995. The National Research Council's Ocean Studies Board received a written request from NOAA (dated January 26, 1994) to conduct a peer review study within six months so as to enable NOAA to use the results for the 1994 ICCAT meeting. Accordingly, the Ocean Studies Board established the Committee to Review Atlantic Bluefin Tuna to review and evaluate the scientific basis of U.S. management of fisheries for Atlantic bluefin tuna and to recommend research to resolve remaining stock structure issues. Members chosen for the committee have a range of expertise, including tuna biology and physiology, fish genetics and stock identification, fish population dynamics, fish ecology, and oceanography (see Appendix A). A notice soliciting comments from the public regarding scientific issues for consideration by the review committee was published in the *Federal Register*.[1] Numerous comments were received, and at the first committee meeting, in May 1994, individuals who had submitted written comments were invited to give a brief presentation to the review committee. The committee met again in June 1994 to hear additional individuals and to complete this report. In addition, the committee reviewed extensive peer-reviewed and gray scientific literature (see Appendix B) as background for its deliberations.

This report focuses primarily on the scientific basis for (1) the assumptions about stock structure and (2) indices of abundances used in the stock assessments for western Atlantic bluefin tuna. The issue of stock structure of Atlantic bluefin tuna is discussed in Chapter 2, and movement of Atlantic bluefin tuna is discussed in Chapter 3. Information on the indices of abundances and the results of sensitivity analyses conducted by the committee using data sets obtained from NOAA's National Marine Fisheries Service and the industry are presented and discussed in Chapter 4. In Chapter 5 the committee summarizes its major findings and makes general recommendations and research recommendations for improving the scientific basis of the management of Atlantic bluefin tuna.

---

[1]*Federal Register*, March 3, 1994, vol. 59, no. 42, pp. 10114-10115.

2

Historical Evidence for Stock Structure

INTRODUCTION

This chapter summarizes available biological evidence on the stock structure of Atlantic bluefin tuna. The chapter reviews published and unpublished information on the issue of whether this information supports the existence of separate eastern and western management units of bluefin tuna in the North Atlantic Ocean. Included are sections on stock concepts, history of stock and population designations in Atlantic bluefin tuna, genetics, life history, climate, and movement. Except for the first two sections (on stock concepts and historical stock designations in Atlantic bluefin tuna), the term "management unit(s)" is used interchangeably with eastern and western stocks of Atlantic bluefin tuna and the term population is used to indicate a genetically-distinct group of fish. In addition, the term "movement" is used to indicate mixing, migration, or both, as they refer to individuals.

CONCEPT OF STOCKS

Geographic boundaries of species are influenced by environment, suitable habitats, and historical events. A fish stock can be defined as all fish belonging to a given species that live in a particular geographic area at a particular time. The area can be constrained by geographic or oceanographic features (e.g., bays, temperature discontinuities), but also may be defined by political boundaries. Political boundaries are commonly used in fisheries management, but a stock defined in this way generally will not reflect biologically meaningful management units.

Criteria for managing fishery stocks under the Endangered Species Act of 1973 (ESA) have been discussed by Waples (1991) and under the Magnuson Fishery Conservation and Management Act of 1976 (MFCMA) in a recent National Research Council report (NRC, 1994). The MFCMA (Public Law 94-265, 16 U.S.C. 1801 et seq.) specifies that "an individual stock of fish shall be managed as a unit throughout its range," but does not provide criteria for defining a stock. The original ESA also did not provide criteria for defining a stock or management unit, but the ESA amendments of 1978 (Public Law 95-632 [1978], 92 Stat. 3751) defined a "species" as "any subspecies of fish or wildlife or plants, and any distinct population segment of any species of vertebrate fish or wildlife which interbreeds when mature." In the past few decades, biochemical and molecular genetic methods have been applied to fishery management issues, leading to an expansion of the stock concept to include interpopulation genetic variability. This expanded stock concept is intended to facilitate the conservation of biologically meaningful management units (Utter, 1981; Waples, 1991) that may be uniquely adapted to a particular area. Waples (1991) proposed that a distinct population segment should be defined as an "evolutionarily significant unit" (ESU) that is "substantially reproductively isolated from other conspecific populations units" and that "represents an important component in the evolutionary legacy of the species." Under this concept, populations are defined as groups of individuals that share a common space, interbreed, and are totally or partially isolated from other such groups. The degree of isolation, brought about by reduced gene flow among breeding areas and the amount of time the populations have been isolated from one another, influences the degree of genetic differentiation among groups.

Dizon et al. (1992) proposed a scheme of population classification based on genetic and geographic criteria. Category I populations are geographically separated groups of individuals that are more closely related genetically to each other than they are to individuals in other groups. Such populations have the highest probability of being evolutionarily significant units. Category II populations are also differentiated genetically but are only marginally separated geographically. Category III populations show little genetic differentiation from one another but are geographically separated and therefore likely to be isolated reproductively. Genetic differences among geographically isolated populations are expected eventually to increase. Category IV populations show little genetic differentiation because of extensive gene flow and are unlikely to be evolutionarily significant units. Geographic distribution, parasite markers, microconstituent analysis, tag-recapture data, population parameters, morphological variability, and genetic information can be used to assign populations to these categories. A variety of evidence is available for assessing the population structure of Atlantic bluefin tuna, but much of the data are equivocal. The sections below present these data and their interpretations.

HISTORY OF ATLANTIC BLUEFIN TUNA STOCK DESIGNATIONS

The two-stock hypothesis currently used by the International Commission for the Conservation of Atlantic Tunas (ICCAT) is based in part on the assumption that mixing of western Atlantic bluefin tuna and eastern Atlantic/Mediterranean bluefin tuna is limited (ICCAT, 1992, 1993). The two management units of Atlantic bluefin tuna include a western management unit (west of 45°W) and an eastern management unit (east of 45°W and in the Mediterranean Sea; see Figure 2-1). Data indicate that while spawning is limited to two discrete areas, the Gulf of Mexico and the Mediterranean Sea, there is movement of individuals between western and eastern management units. A key issue is the extent of movement.

The first studies of population structure in Atlantic bluefin tuna date to the early part of this century. The earliest written work is attributed to several reports by M. Sella (1926, 1927, 1929; cited in Brunenmeister, 1980), who, in the mid- to late 1920s, inferred origins and movement patterns from fishing tackle characteristics of different eastern Atlantic Ocean and Mediterranean Sea fisheries. Sella hypothesized that bluefin tuna moved from the eastern Atlantic Ocean into the Mediterranean Sea, that tuna moved from the south of Spain to Norway after spawning, and that small and medium-sized tuna could swim long distances. Although criticized because of concerns that tackle types were not reliable indicators of hooking localities, Sella's hypotheses agree with movement patterns inferred from tagging experiments carried out since the early 1900s.

An early review of research on population structure for Atlantic bluefin tuna is an unpublished report in the early 1970s by F.J. Mather and A.C. Jones (cited in Murphy, 1990), which suggested that there were three populations: one in the western Atlantic Ocean, one in the eastern Atlantic Ocean, and one in the Mediterranean Sea. They also suggested that a separate population might exist in the south Atlantic Ocean. Mather et al. (1974) suggested two alternative hypotheses: a single Atlantic population and one or more Mediterranean populations; or two Atlantic populations, one spawning in the western Atlantic Ocean and the other in the eastern Atlantic Ocean or the Mediterranean Sea, or both, and one or more Mediterranean populations. They believed there was evidence of a two-stock hypothesis but noted that the "evidence is insufficient to permit clear-cut conclusions." Brunenmeister (1980) also reviewed evidence for population structure but was unable to support any hypothesis. Finally, Murphy (1990) argued that the presently accepted two-population hypothesis was not adequately flexible to fit available data. He proposed that bluefin tuna in the northern Atlantic Ocean represented one population and that the interchange between the Atlantic and Mediterranean populations is sufficiently small such that Mediterranean bluefin tuna may represent a second population. His hypothesis appears to be based on three assumptions: (1) rates of movement between western and eastern Atlantic bluefin tuna are significantly higher than those between eastern Atlantic and

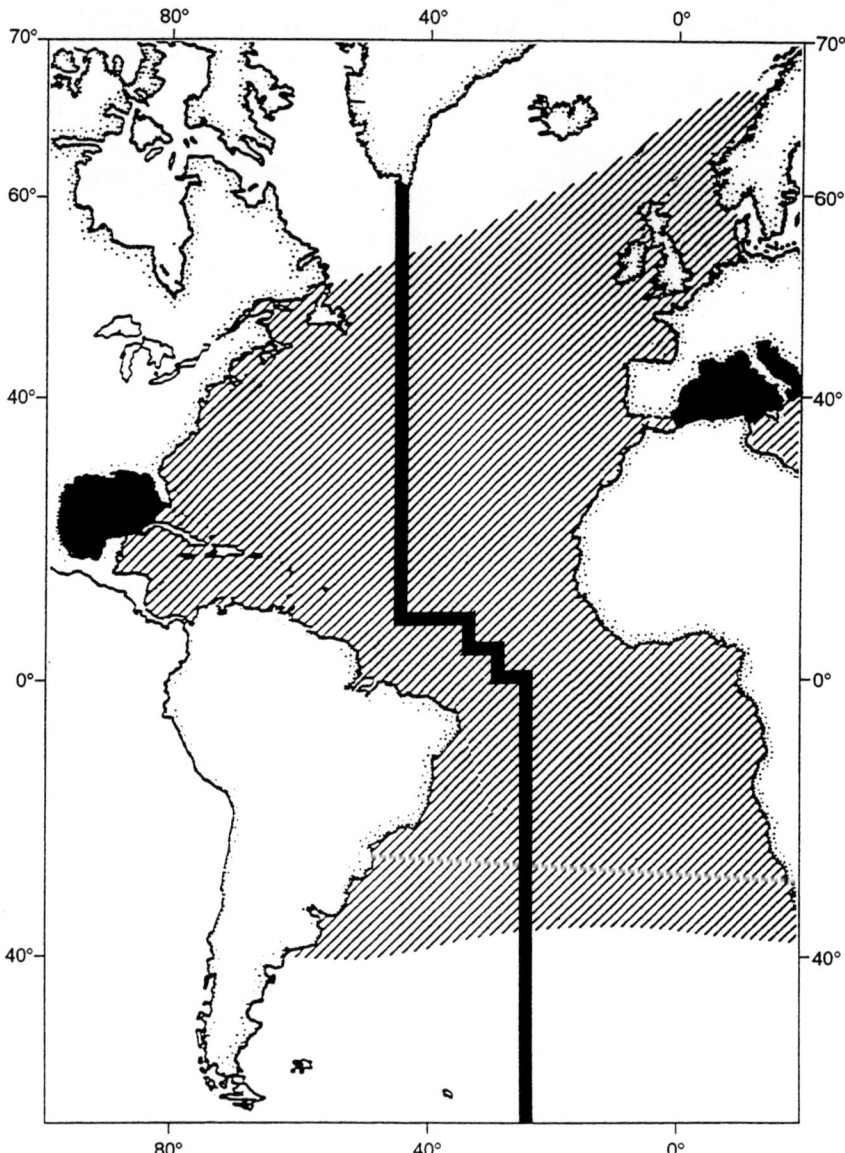

FIGURE 2-1 General distribution of bluefin tuna in the Atlantic Ocean (darkened areas indicate known spawning areas [adapted from FAO, 1968]). The solid line separates the ICCAT eastern and western management units.

Mediterranean bluefin tuna; (2) possible differences in spawning productivity exist; and (3) the Gulf Stream, via its extension in the North Atlantic Current, could transport larval and postlarval bluefin tuna into the middle of the Atlantic Ocean.

GENETIC STUDIES

Molecular techniques that assay genetic variability have an obvious advantage over techniques that measure life history traits where the genetic component is unknown. Such techniques can be used to quantify genetic divergence and gene flow among populations and to estimate breeding structure within populations. There are four criteria for using a molecular method to find genetic markers that may be used to define fish management units: (1) expression of the genetic markers does not change during the life of an individual; (2) barring mutation, markers are inherited unchanged from one generation to the next; (3) it is possible to assay a large number of individuals from a large number of localities to adequately resolve the genetic structure of one population; and (4) there is sufficient within-population variability to make make robust statistical tests of geographic structure.

Data resulting from methods satisfying these criteria can be analyzed in several ways. If genotypic data from nuclear genes (allozymes) of randomly sampled individuals surveyed with protein electrophoresis are available, a contingency table analysis of gene frequencies can be used to test for homogeneity among different sampling localities. If significant differences are found, and one can assume migration-drift equilibrium, then one can infer that the samples were drawn from genetically discrete populations. A caveat is that large samples must be used to detect small but significant gene-frequency differences between or among areas. Another test for geographic structure is to compare the observed numbers of genotypes (*AA, AB, BB,* etc.) in a pooled sample with the number expected from random mating (Hardy-Weinberg proportions),

$$
\begin{array}{ccc}
AA & AB & BB \\
p^2 & 2pq & q^2
\end{array}
$$

where p is the frequency of the A allele and q is the frequency of the B allele. If there are regional genetic differences, the pooled sample will show a significant deficit of heterozygotes (Wahlund's effect) owing to mixing of individuals from discrete populations. This test, however, lacks power to detect small but significant genetic differences between populations. This significant deficit is taken as evidence for genetic differentiation between or among populations.

More recently, methods for detecting nucleotide sequence variability have been used to study fish populations. Most of this effort has been directed toward the analysis of mitochondrial DNA (mtDNA), a circular piece of DNA found in the cytoplasm of the mitochondria outside the cell nucleus. The analysis of mtDNA is based on three important differences from the analysis of nuclear DNA. First, mitochondria are inherited only from the female parent; second, there is no recombination among mtDNA molecules, greatly simplifying interpretation of phylogenetic trees based on mtDNA; and, third, there is a higher mutation rate in mtDNA than in nuclear genes typically assayed for population discrimination, thus providing the opportunity to examine more recent divergences. These characteristics provide a finer-scale resolution of genetic differences than is possible for the analysis of nuclear genes assayed with protein electrophoresis. The rates of mtDNA evolution in some large animals, however, appear to be quite slow and the analysis of mtDNA in these animals may not provide better population resolution (e.g., skipjack tuna; see Appendix E). DNA restriction (cleaving) enzymes have been used extensively to detect mtDNA restriction site differences among haplotypes,[1] and, more recently, the polymerase chain reaction (PCR) has been used to amplify specific mtDNA segments for nucleotide sequencing. One application of mtDNA restriction site or sequence information (or both) is the generation of parsimony networks for visualizing minimal mutation distances among haplotypes. These networks link mtDNA haplotypes by single gains or losses of restriction sites (or by changes in homologous nucleotides) and are superimposed on geographic localities to test whether geographic cohesion of haplotypes (or haplotype lineages) exists. If haplotypes are shared among localities, frequency distributions also can be used in contingency table or other analyses to test hypotheses of genetic homogeneity among geographic localities.

The level of population structure that can be detected with a particular molecular or biochemical method depends on the mutation rate associated with the assayed DNA or its product. Techniques such as protein electrophoresis, which distinguish products of genes with mutation rates of about 10^{-7} per generation (Nei, 1987), are capable of detecting the effects of population events over several thousands or millions of years ago. Analysis of mtDNA, which has a mutation rate of about 10^{-5} per generation (Wilson et al., 1985), can potentially detect the effects of more recent population events. Recently, analysis of nuclear microsatellite loci, which appear to have a mutation rate on the order of 10^{-3} per generation (Valdes et al., 1993), has been used to detect events occurring on the order of thousands of years ago. Interpretation of data derived from these various methods should be tempered with an understanding of the temporal scale of population events that each method is capable, in theory, of resolving.

---

[1]The term haplotype refers to the mtDNA genes which are inherited only from the female parent, so the mtDNA carries only a single copy of each mitochondrial gene.

Genetic Variation in Tunas and Scombroid[2] Fish

Bluefin tuna are members of the family Scombridae, an assemblage that contains 15 genera and about 48 species of epipelagic fish. The scombrid genus *Thunnus* contains seven species, including the bluefin tuna. Results of molecular genetic studies of *Thunnus*, as well as species that are related to bluefin tunas (mackerel, bonitos, and billfishes), have provided information on the amount of genetic differentiation that might be expected among the global populations of bluefin tuna. Several recent studies have used mtDNA sequence variation to examine patterns of molecular genetic divergence and infer evolutionary relationships among tunas and other scombroid fish. Bartlett and Davidson (1991) examined a 290-base-pair (bp) sequence of the cytochrome *b* (cyt *b*) gene in four species of tuna from the northeast Atlantic Ocean: bluefin ($n=33$), yellowfin ($n=33$), bigeye ($n=32$), and albacore ($n=12$). The main result of this study was the demonstration that one could use these 290 bp to distinguish each of the four species and that only a small amount of tissue (100 mg) is required for analysis. This was significant given that carcasses (which do not have identifying morphological characters), often end up at auctions and could be sampled and used in the genetic assessment of population structure. Also, identification of larval and juvenile specimens, which are often difficult to sort to species, is now possible with genetic techniques.

In broader studies to understand the relationships among tunas, mackerels, and billfishes, Block et al. (1993) and Finnerty and Block (1994) examined 600 nucleotides of the cyt *b* gene among 30 species. These studies included nine species of tunas and provided direct comparison of sequence variability between northern and southern bluefin tunas. The inferred phylogeny provided strong support for the monophyly of tuna genera (*Thunnus*, *Katsuwonus*, *Euthynnus*, and *Auxis*) and indicated that species of the genus *Thunnus* are closely related to one another. Only three nucleotide sites differed between southern ($n=2$) and northern ($n=2$) bluefin tuna. Sequence differences between these populations are small (0.5%) compared to the maximum intraspecific sequence difference detected among other members of the suborder (i.e., 1.8% for blue marlin, *Makaira nigricans*, over a similar region of cyt *b* [Finnerty and Block, 1992]). These data call into question the validity of separating northern (*T. thynnus*) and southern bluefin tuna (*T. maccoyi*) into separate species or indicate the presence of Northern bluefin tuna in southern oceans. Additional molecular data, particularly from nuclear genes, are needed to determine whether the inferences made in these studies can be corroborated.

In a third study of mtDNA variability in tunas, Chow and Inoue (1993)

---

[2]The suborder that includes tunas and other fish is Scombroidei, and these fish are referred to as scombroids. When referring specifically to tuna, the family name, scombridae, is used, and these fish are referred to as scombrids.

examined restriction fragment length polymorphisms (RFLPs) of three PCR-amplified mtDNA fragments from eight species/subspecies of *Thunnus*: (1) cyt *b* (355 bp, 15 enzymes, $n = 132$); (2) 12S ribosomal RNA (450 bp, 20 enzymes, $n = 16$); and (3) part of the coding regions of the ATPase and COIII genes (ATCO, 940 bp, 20 enzymes, $n = 131$), as well as regions adjacent to the coding regions. Cyt *b* RFLP fragment differences separated these tunas into four groups. The groupings identified from the cyt *b* data by Chow and Inoue (1993) were discordant with groupings inferred from direct sequencing of cyt *b* (Block et al., 1993) and from analyses of morphological variation (Collette et al., 1984). Chow and Inoue (1993) also reported that ATCO fragments indicated differences among species. One of the 18 northern Pacific bluefin tuna had cyt *b* and ATCO fragment patterns identical to those of northern Atlantic bluefin tuna. The authors suggested that this reflects incomplete genetic differentiation between northern Atlantic and Pacific bluefin tuna. A more reasonable hypothesis is that the exceptional mtDNA haplotype reflects movement of Atlantic bluefin tuna into the Pacific Ocean basin. The possibility of movement of bluefin tuna between Atlantic and Pacific ocean basins merits further investigation.

Genetic Variation in Bluefin Tuna

Information on biochemical and molecular genetics of Atlantic (*Thunnus thynnus*) and southern (*T. maccoyi*) bluefin tuna populations is limited. One early study of frequencies of alleles coding for the protein transferrin among four samples of southern bluefin tuna in Australian waters showed significant allele-frequency differences ($G = 21.02$, degrees of freedom = 3, $P < 0.001$; Fujino and Kang 1968). In Atlantic bluefin tuna, Edmunds and Sammons (1971, 1973) found allele-frequency homogeneity at the superoxide dismutase (SOD) locus among samples from the western Atlantic Ocean (New Jersey, $n = 269$; Rhode Island, $n = 87$; and Nova Scotia, $n = 25$) and between these samples and one from the Bay of Biscay ($n = 675$) in the eastern Atlantic Ocean (G-test, $P > 0.05$). Edmunds and Sammons (1973) pooled these samples into a single sample ($n = 1,056$) and tested the pooled sample for fit to Hardy-Weinberg proportions. No deviation from expected equilibrium proportions was found. This was taken as evidence for the lack of genetic differentiation between eastern and western Atlantic bluefin tuna. The oceanwide fit of SOD genotypes to proportions expected from random mating is consistent with the hypothesis that a single population of bluefin tuna occurs in the North Atlantic Ocean.

Using allele frequencies at three nuclear gene loci, Phipps (1980) tested the hypothesis that early- and late-arriving bluefin tuna in St. Margaret's Bay, Canada, did not differ genetically. Most bluefin tuna enter the bay in two waves, one in July and a smaller one in mid-September/October (429 were sampled in the first wave, 16 in the second wave). No departures from expected Hardy-Weinberg proportions within or among samples were found at SOD-1, SOD-2, and G_6PD (glucose-6-phosphate dehydrogenase). Allele frequencies at SOD-1 between the two samples did not differ significantly, whereas allele frequencies at SOD-2 ($G = 12.11$, $P < 0.01$) and G_6PD ($G = 137.19$, $P < 0.001$) did differ significantly. There was evidence for a null allele at SOD and for artifacts of G_6PD banding, indicating that the differences between samples must be interpreted with caution. In addition, the sample of late-arriving fishes was small (n=16).

Bartlett and Davidson (1991) sequenced 290 bp of the cyt *b* gene and found six mtDNA haplotypes among 33 individuals of Atlantic bluefin tuna; 28 of the fish shared a single haplotype. The common haplotype differed from four others by one nucleotide substitution and from one other by two substitutions. Because the portion of the cyt *b* gene sequenced by Bartlett and Davidson (1991) generally is conserved in other scombroids (e.g., swordfish [Finnerty and Block, 1992]), it likely will not be informative for resolving population structure in Atlantic bluefin tuna. Other regions of the mtDNA molecule (e.g., the D-loop) are more variable and thus may be more suitable for resolving population structure in bluefin tuna.

To date, molecular genetic studies of bluefin tuna have not focused on the issue of genetic divergence among global samples of bluefin tuna. The genetic analysis of within-ocean basin diversity of Atlantic bluefin tuna (e.g., Atlantic Ocean, Mediterranean Sea), would benefit from a worldwide study of molecular genetic variation among bluefin tuna. Molecular genetic studies in other highly migratory, scombroid fishes (tunas, marlins, and swordfish) have demonstrated the utility of such an approach (Appendix C), and a thorough analysis of nucleotide sequence variability in both mtDNA and rapidly evolving nuclear DNA (micro- and/or minisatellite loci) in Atlantic bluefin tuna should be encouraged.

Conclusion

There is less genetic information available for Atlantic bluefin tuna than for other scombroid fish. The studies of Edmunds and Sammons (1971, 1973) are consistent with the hypothesis that eastern and western management units of Atlantic bluefin tuna comprise a unit Mendelian population (i.e., they are genetically homogeneous). The remaining studies are either incomplete or inadequate to address the issue.

Recommendation

A major research effort should be undertaken to thoroughly assess the genetic basis of the population structure of Atlantic bluefin tuna. Multiple genetic characters, detected by a variety of approaches, should be employed to provide information on several fundamental questions and to resolve the issue of stock structure. It is critical to support a variety of genetic studies.

LIFE HISTORY PARAMETERS

This section focuses on aspects of life history of Atlantic bluefin tuna that may be relevant to management. All known aspects of bluefin tuna life history have been discussed in detail by Clay (1990) and will not be reviewed here. The committee holds the view that there are aspects of life history that may influence catch per unit effort (CPUE) or other indices used in stock assessment models. We also argue that there are important aspects of life history that are not considered in ICCAT's Standing Committee on Research and Statistics (SCRS) deliberations. Further, much of the life history of Atlantic bluefin tuna is not known.

Life history variables such as age composition, growth, age at maturity, and mortality have been used to infer the population structure of several fish (Ihssen et al., 1981). When used as evidence for two populations, these measures are indirect indicators of possible genetic differences. They also can reflect individual responses to environmental differences among localities, so that conclusions concerning population structure based on these data are suspect since they cannot differentiate between genetic and environmental influences. If differences in life history variables result entirely from environmental factors, the two-population hypothesis cannot be tested with these values. In theory, natural selection, genetic drift, and migration between or among localities determine the degree that life history variables change from one locality to the next: high levels of migration tend to minimize differences among localities, whereas extremely low levels of migration could allow population differences to appear in only a few generations.

Geographic Locality of Spawning Grounds

Although bluefin tuna have been found as far north as Newfoundland in the western Atlantic Ocean and as far north as Norway in the eastern Atlantic Ocean, and a fishery existed for a short time as far south as Brazil, extensive searching has detected only two spawning localities: the Gulf of Mexico and the Mediterranean Sea (Figure 2-1). Each of these localities is large relative to the spawning areas of many other fish species, but small relative to the spawning areas of tropical tunas. Individual females in both the east and the west produce about 30,000,000 eggs each (Clay, 1990). There is no evidence that the large geographic separation of the spawning localities represents reproductive separation.

Richards has reviewed evidence of spawning in the Gulf of Mexico (Richards, 1976) and summarized results of ichthyoplankton surveys in the western Atlantic Ocean (Richards, 1987). Larvae and juveniles are found primarily in the northern region of the Gulf of Mexico, with sporadic occurrences in the Florida Straits and off the Texas coast. Larvae have been sampled off the Carolina coast in the western Atlantic Ocean, but their presence there may result from advection by currents from the Florida Straits and not from local spawning

(Richards, 1990). No bluefin tuna larvae have been found in the eastern Atlantic Ocean (Cort and Loirzou, 1990a), and it has been assumed that bluefin tuna do not spawn there.

The western Mediterranean Sea and the Adriatic Sea appear to constitute the second major spawning location for bluefin tuna, but the highest concentrations of bluefin larvae are found in the central part of the western basin of the Mediterranean Sea between southern Italy and Sardinia and around the Balearic Islands of Spain (Cort and Loirzou, 1990a).

There are some key questions regarding these spawning locations. What are the biological requirements for spawning, and what environmental factors trigger spawning? How much spawning occurs in one location relative to another, and does it vary from year to year because of changes in environmental conditions? A scientific effort should be made to learn the biological and environmental requirements of spawning. Some scientific effort should also be directed toward estimating the relative amounts of spawning in the two locations, by using identical survey methods in both locations.

Timing of Spawning

Most aspects of spawning in Atlantic bluefin tuna are still unknown because spawning has not been observed. It is not known if bluefin tuna spawn once or many times per season or whether an individual spawns yearly. It is also not known whether individuals can spawn in the east and then in the west at different times. Spawning in the Gulf of Mexico reportedly occurs from mid-April to mid-June (Richards, 1990; Dicenta et al., 1980). Spawning in the Mediterranean Sea is thought to occur from June to August (Rodriguez-Roda, 1971; Dicenta et al., 1980; Cort and Loirzou, 1990b). It is not known, however, whether later spawning times for the Mediterranean fish are based on genetic differentiation or whether they are in response to environmental differences between the two locations. Differences in spawning times do not necessarily indicate that bluefin tuna produced in the Gulf of Mexico and maturing there or in the western Atlantic Ocean would be precluded from spawning, as adults, in the Mediterranean Sea, or vice versa. Tagging experiments demonstrated that fish can cross the Atlantic Ocean in less than 60 days. It is possible for a fish to spawn in the west in April, migrate to the east, and arrive in time to spawn in the east the same year.

Earlier studies providing information on water temperature during spawning of bluefin tuna are referred to in Tiews' (1963) review of biological data on bluefin tuna. A range of 24.9°C to 29.5°C is reported for the Straits of Florida, from Havana to Bimini (Rivas 1954). In the central Mediterranean Sea, the reported range is 19°C to 21.6°C (Roule 1924). Also, for the Mediterranean Sea, large and small sized, sexually mature bluefin tuna spawn at different water temperatures, the larger fish in water ranging from 18°C to 22°C and the smaller fish in warmer waters ranging from 23°C to 25°C (Sella, 1931).

Age at Sexual Maturity

Baglin (1982) found no mature fish among 12 six-year-old females and 15 seven-year-old females sampled during June in the Mid-Atlantic Bight, whereas generally it is accepted that eastern Atlantic/Mediterranean bluefin tuna mature sexually in the third year (Tiews, 1963; Rodriguez-Roda, 1971; Cort and Liorzou, 1990b). The study by Rodriguez-Roda (1967) of 50 eastern Atlantic/Mediterranean bluefin tuna en route to spawning locations appears to be the primary data for the postulated size at maturity in this stock: 50% of three- to four-year-old fish were mature, whereas 100% of four- to five-year-old fish were mature (Cort and Loirzou, 1990b). The works of Baglin (1982) and Rodriguez-Roda (1967) appear to form the primary basis for the hypothesis that size at, and age of, sexual maturity in western Atlantic bluefin tuna (200 cm and 10 years) are larger and greater, respectively, than for eastern Atlantic/Mediterranean bluefin tuna (130 cm and five years).

Clay (1990) criticized Baglin (1982), noting that (1) sample sizes of each year class in Baglin's study were small; (2) sampling was carried out in June, near the end of the spawning season; and (3) vitellogenic oocytes in the six- and seven-year-old fish were being resorbed. This note concerning resorption is important because rapid resorption in fish makes it impossible to report whether or not a particular fish has spawned if it has been sampled only a few weeks after the spawning period (VanDerKraak, pers. comm.).[3] On this basis, Clay (1990) suggested that western Atlantic bluefin tuna also might become sexually mature before the age of 10 years. Clay (1990) also criticized the study of Rodriguez-Roda (1967) by noting that only one immature female in the length interval of 75 to 80 cm was included in his analysis (the next size interval was 115 to 120 cm, where all fish were mature) and that the fish were from the south Atlantic coast of Spain and may not have adequately represented the eastern Atlantic/Mediterranean population. Clay (1990) noted that the sample sizes and seasonal representation of individuals in both studies were inadequate and that conclusions based on the hypothesis of differences in spawning locations between western Atlantic and eastern Atlantic/Mediterranean bluefin tuna were compromised. The geographic extent of sampling in these studies was inadequate. The basis for a difference in spawning age and size between western Atlantic and eastern Atlantic/Mediterranean bluefin tuna depends primarily on two studies, neither with adequate sampling of seasons and locations. Because differences in life history characteristics could signal significant biological or genetic differences among

---

[3]VanDerKraak, G. July 1994. Department of Zoology, University of Guelph, Guelph, Ontario, N1G 2W1, Canada.

areas, additional studies of age and size at spawning should be carried out. In particular, the committee suggests that all fish caught as by-catch in the Gulf of Mexico be sampled: the gonads should be taken for histological analysis, and body length should be measured to estimate age. It is more important to sample males than females because females have a higher rate of resorption than males, making their data a less reliable indicator of recent spawning history.

Larval Biology

Bluefin tuna larvae grow rapidly (about 1 mm per day), and larval abundance in spawning locations ranges from 0.1 to 1.0 per square meter (Clay, 1990). Little information exists regarding food requirements, food availability, and potential predators of larvae, and it is not known whether larval growth is limited by food availability. There is little information about the movements of bluefin tuna larvae.

Physiological Ecology of Bluefin Tuna Movement Patterns

Numerous reports and papers on bluefin tuna describe the species as highly migratory. Figure 2-1 shows the distribution of bluefin tuna in the Atlantic

TABLE 2-1 Migration speed of Atlantic bluefin tuna (kilometers/day) calculated from tag-recovery data (Clay, 1990). Only speeds greater than 1 nautical mile/day are reported, and all values represent minimum estimates.

	Age 0 Italy	Age 0 Spain	Age 1 Morocco	Age 1 France	>150cm Norway	>150cm Spain
	9.7	1.9	3.2	2.5	9.8	150.0
	9.7	5.7	3.6	18.9	9.9	9.9
	12.4	3.7	2.4	13.2	9.4	15.7
	2.0	6.8	2.1	—	10.8	3.2
	—	3.7	4.4	—	14.8	48.2
	—	5.9	—	—	8.0	34.4
	—	6.2	—	—	10.7	96.3
	—	—	—	—	10.0	17.5
	—	—	—	—	2.0	5.2
	—	—	—	—	8.5	32.1
	—	—	—	—	9.8	20.9
	—	—	—	—	4.0	—
Mean	6.7	4.2	2.6	8.6	8.3	36.1

Mean for age 0 and age 1 fish = 6 km/day

Mean for giant fish = 24 km/day

Ocean. Table 2-1 presents data based on tag recoveries and shows that minimum estimates of migration speed are modest. Age 0 and age 1 fish can move about 6 km/day, and large tuna (>150 cm) can move about 24 km/day.

Clay's 1990 review of migration patterns is a recapitulation of a model by Rivas (1978). The proposed migratory patterns suggested by Rivas are consistent with the limited tag-recovery data. It should be noted that Rivas entitled his paper "Preliminary Models," but the tendency has been to accept his preliminary models as fact. Rivas' models are consistent with what is known about bluefin tuna movements, but there is a great deal that we do not know.

In the following section, what is known about the movement patterns will be discussed from a functional point of view; the functions being reproduction and moving to locations of high food density. Spawning locations probably are favorable environments for the young but unfavorable environments for giant tuna[4] because of high ambient temperatures. Tunas have a high rate of heat production (relative to nontunas) and a well-developed heat conservation system. Their metabolic rate has not been measured, but based on surface area and structure of the gills and on metabolic capacity of the skeletal muscles, it is likely that bluefin tuna have a high rate of heat production. The few available measurements of muscle temperature do not allow us to conclude that bluefin tuna can easily remove the heat that is produced. For example, sportfishermen report that giant fish tire more quickly and are more easily landed in warm than cold water; this suggests that high metabolic rate and high ambient temperture act synergistically to cause distress. Observations suggest that the warm waters of the Gulf of Mexico may be unfavorable for giant fish. Western Atlantic bluefin tuna occur in waters as cold as 6.6°C (Squire, 1962), whereas eastern Atlantic bluefin tuna have not been reported in waters cooler than 12°C (Tiews, 1962), and their migration into and from the North Sea was closely associated with movement of the 12°C isotherm (Luhmann, 1959).

In both the east and the west, there are large-scale movements of giant fish toward spawning grounds, but the pattern of movements to these spawning locations and from where they come is poorly known. It is reasonable to conclude that spawning locations are preferred because they provide a favorable environment for eggs and larvae. That is, they provide abundant food to support the rapid growth of the young and are locations of relatively low predation.

Movements of giant fish away from spawning grounds are better known than those to the spawning grounds. The movements serve two functions: move to cooler water, and move to locations of high food density. The limited evidence available suggests that the migrations are at a steady speed and are fast relative to other species. Bluefin tuna migrate from the Gulf of Mexico northward along the U.S. coast to the Canadian coast. When they leave the spawning

---

[4]Giant fish are described as greater than 180cm and greater than 130kg (Clay 1990).

ground, they have a low lipid content and thus are of low value to commercial fisheries, as the price of tuna increases markedly with an increase in lipid content.

During this migration, the tuna encounter patches of food. In general, they move in a northerly direction during this stage of migration and gain weight, mostly as stored lipid. Water temperature is thought to exert a strong influence the pattern of movement during searches for new patches of food.

This aspect of life history is important to ICCAT assessment procedures for two reasons. First, local tuna abundance depends markedly on food availability. It seems appropriate to estimate the availability of food such as squid, herring, and mackerel in the various regions of the western Atlantic Ocean. This information should not be used as a bluefin tuna index, per se, but might be useful in choosing or weighting bluefin tuna abundance indices in the virtual population analyses (VPA)[5] or in designing various sampling schemes. Second, sampling for any CPUE index must be of a scale that takes into account the scale of food patches. The importance of patchiness probably has been underestimated in the biology of bluefin tuna and in the assessment of their abundance.

Describing bluefin tuna as highly migratory is accurate but not complete. Examination of catch records or CPUE data clearly shows that the distribution of bluefin tuna is patchy, and the scale of patchiness is large in time and in space. The ICCAT model relies on a temporal series of CPUE data. If the spatial scale of patchiness at one time is large relative to the scale of the sampling, the usefulness of the index will be compromised.

For example, there are large-scale changes in core rings of the Gulf Stream that cause changes in all levels of the food chain and are expected to alter movement patterns of bluefin tuna as they search for food. The spatial scale of these rings is rather large. One would predict changes in local bluefin tuna abundance over distances of up to 100 km, and this is concordant with some of the catch data. Table 2-2 and Figure 2-2 show catch data for Canadian waters. Figure 2-3 is a map of the location of the Canadian catch data presented in Table 2-2 and Figure 2-2. Examination of catches at any one location (i.e., reading down one data column or across one figure panel) shows large changes with a period of several years and may lead to the conclusion that the stock is crashing. The "crashes," however, do not occur simultaneously. The earliest crash was at Wedgeport, Nova Scotia (data not shown in Table 2-2 or Figure 2-2). The Sharp Cup was the major international tuna tournament in the 1930s. The number of

---

[5] The virtual population analysis (VPA) model used in fisheries analysis is based on a simple concept: if the annual catch in numbers of each year class and the corresponding natural and fishery mortality can be estimated, the numbers of each year class in the populations can be calculated for each year and the population reconstructed on paper. It is useful in testing the validity of estimates of recruitment and mortality (Wise, 1991).

TABLE 2-2 Landings of giant Atlantic bluefin tuna by numbers of fish from Canadian fisheries by province from 1960 to 1989[1] (adapted from Clay and Hurlbut, 1989, and Clay[2] personal communication, 1994).

Year	1 N.S. trap	2 N.S. in	3 N.S. long	4 N.S. off	5 N.S. Gulf	6 P.E.I. Gulf	7 N.B. Gulf	8 Que. Gulf	9 Nfld. in	10 Nfld. long	11 Nfld. off	12 Total
1960	—	—	—	—	—	—	—	—	11	—	—	11
1961	—	—	—	—	—	—	—	—	133	—	—	133
1962	—	—	—	—	—	—	—	—	—	—	—	0
1963	—	—	—	—	—	—	—	—	—	—	—	0
1964	—	—	—	—	—	—	—	—	—	—	—	0
1965	286	73	—	—	—	—	—	—	223	—	—	582
1966	306	30	—	—	—	—	—	—	388	—	—	724
1967	614	23	—	—	—	5	—	—	179	—	—	821
1968	356	53	—	—	—	13	—	—	604	—	—	1,026
1969	680	12	—	—	—	31	—	—	585	—	—	1,308
1970	458	15	—	—	—	99	—	—	418	—	—	990
1971	208	9	—	—	—	173	—	—	76	—	—	466
1972	104	12	—	—	—	482	—	—	157	—	—	755
1973	508	19	—	—	—	653	4	—	37	—	—	1,221
1974	865	—	—	—	22	1,048	93	6	30	—	—	2,064
1975	452	—	—	—	10	343	148	6	33	—	—	992
1976	474	—	—	—	—	650	180	26	6	—	—	1,336
1977	948	—	—	—	13	448	196	95	5	—	—	1,705
1978	530	—	—	—	17	437	35	11	2	—	—	1,032

Year	1	2	3	4	5	6	7	8	9	10	11	12
1979	72	—	—	111	317	55	20	1	—	—	—	576
1980	129	—	—	50	389	118	90	1	—	—	—	777
1981	93	—	—	81	515	26	29	3	—	—	—	747
1982	157	—	—	61	392	53	43	7	—	—	—	713
1983	17	—	—	20	789	125	54	3	—	—	—	1,008
1984	8	—	—	100	384	78	17	3	—	—	—	590
1985	27	—	—	20	221	47	11	4	—	—	—	330
1986	2	343	—	8	75	2	5	5	—	—	—	440
1987	47	184	—	28	55	1	1	4	148	—	—	468
1988	43	853	988	10	119	—	—	7	637	—	131	2,788
1989	3	287	1,760	24	71	7	—	17	373	—	578	3,120

Note: Numbers exclude fish tagged and released, and exclude fish from the Canadian purse seine fishery off the New England coast. Column 1 is the trap fishery in St. Margaret's Bay, N. S. Column 2 is the rod-and-reel fishery from Shelbourne to Canso off the N.S. west shore. Column 3 is the longline training at the edge of shelf off N.S. Column 4 is the recent offshore longline training fishery in the Hell Hole between Browns Bank and Georges Bank. Columns 5, 6, 7, and 8 are for the rod-and-reel (and tended-line after 1981) fishery in the Gulf of St. Lawrence. Column 9 is the NewFoundland inshore fishery centered on Conception Bay, later spreading to Trinity Bay and Notre Dame Bay. Column 10 is the longline training at the edge of the shelf south of Nfld. Column 11 is the recent offshore longline training fishery at Virgin Rocks. Column 12 is the total yearly landings from columns 1-11 (N.S.=Nova Scotia, P.E.I.=Prince Edward Island, N.B.=New Brunswick, Que.=Quebec, and Nfld.=NewFoundland). Dash indicates no known reported landings.

[1]This was the last year in which Canadian landing data was available for the Committee's analysis.
[2]Douglas Clay, Marine and Anadromous Fish Division, Gulf Fisheries Center, Department of Fisheries and Oceans, P.O. Box 5030, Moncton, New Brunswick, CANADA E1C 9B6.

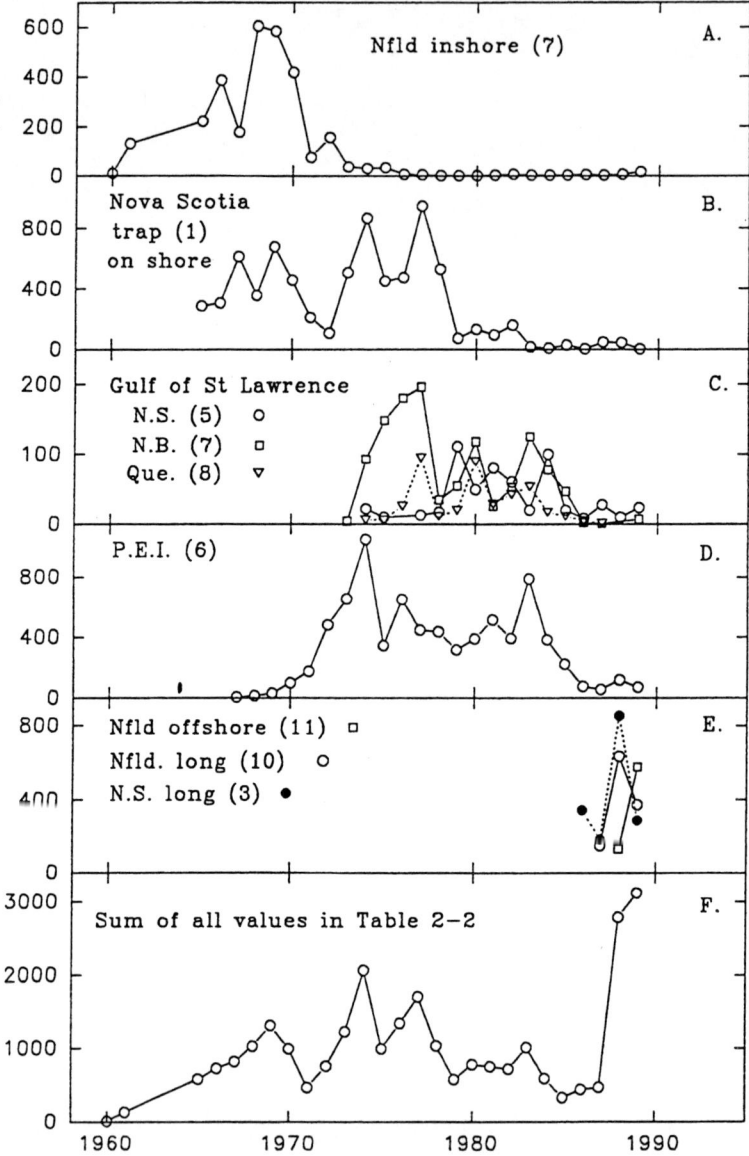

FIGURE 2-2 Landings of giant Atlantic bluefin tuna (by numbers of fish) in Canadian waters showing variation in catch from year to year. Data are listed in Table 2-2. Numbers indicated in panels A-E correspond to the areas listed in the Table 2-2 column headings. Figure 2-3 shows the localities of the Canadian catch data. Note: The values plotted in Panel F are the total Canadian landings listed in Table 2-2, column 12 and are not just the sum of the landing in panels A-E.

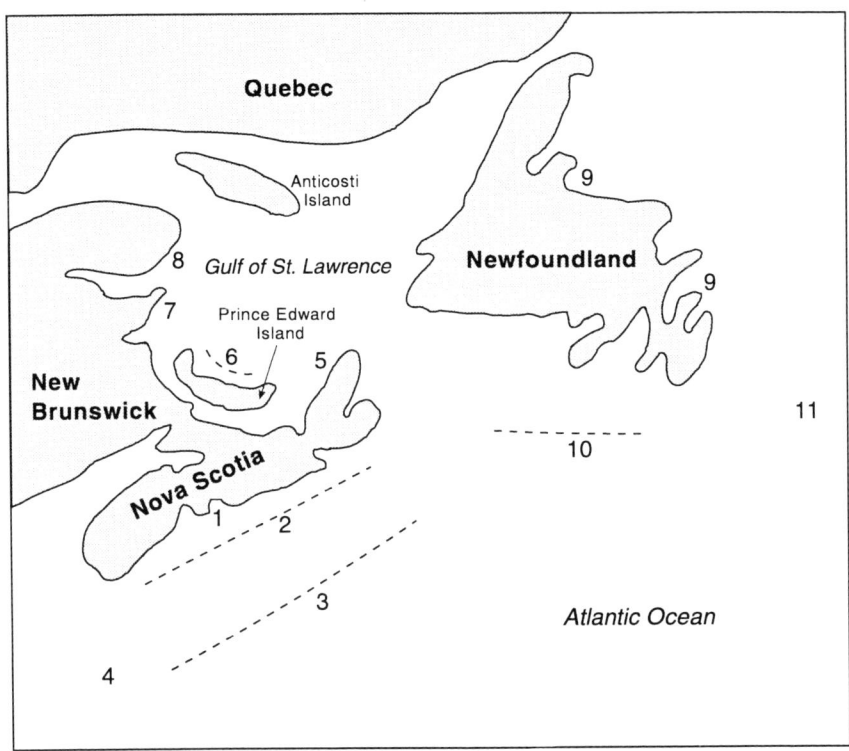

FIGURE 2-3 Map showing the localities of the Canadian catches from Gulf of St. Lawrence and surrounding Atlantic Ocean regions presented in Figure 2-2. Numbers correspond to the areas listed in Table 2-2 column headings.

giant fish that landed annually increased from 384 in 1946 to a peak of 1,760 in 1949. No fish were caught in 1966. Thus, the apparent crash at this location occurred much before any of the others (shown in Table 2-2 or Figure 2-2). Panel E in Figure 2-2 shows the recent development of the offshore fisheries. Portions of this fishery occur in an area close to that of the Wedgeport tournament, which collapsed in the 1950s. Panel F in Figure 2-2 is a plot of the total Canadian landings listed in Table 2-2 and not just those listed in panels A - E. There are two reasonable interpretations of these data and for similar data from other locations:

1. Depletion in a particular location represents the exploitation of a discrete assemblage of fish such as a school or a group of schools returning to a specific feeding area. That is, certain substocks may return to a particular location to feed and these *individual* substocks can be fished out (Clay and Hurlbut, 1989).

2. Depletion in any particular location reflects large-scale patchiness (in time and space) of bluefin tuna abundance and is related to environmental shifts or prey species also responding to environmental changes or both.

One of the columns in Table 2-2 (column 6, P.E.I. Gulf) is used as an index in the VPA model. Given either of the above interpretations of the data, the inclusion of the Prince Edward Island (P.E.I.) tended-line data in the VPA model seems tenuous, because the decline of abundance in this area may not represent abundances at other locations.

Giant Bluefin Tuna in Winter

Little is known about movements after the autumn feeding along the eastern coast of the United States and Canada. The success of a longline fishery in the mid-Atlantic Ocean region suggests movement from the northwest Atlantic Ocean to the mid-Atlantic Ocean region, or from the northeast Atlantic Ocean to the mid-Atlantic Ocean region. During the U.S. exploratory fishing trials from 1957 to 1965, giant bluefin tuna were caught in the mid-Atlantic Ocean region in May but were not caught in January and February (Wilson and Bartlett, 1967). The mid-Atlantic fishery apparently catches fish that moved into this area since the 1960s, perhaps related to changes in the ocean environment.

Movements of Age 0 Fish, Small Fish, and Medium Fish[6]

These movements are driven by searches for water with favorable temperature and food rather than by reproduction. Age 0 fish move from the Gulf of Mexico as far north as Cape Cod. "Small" bluefin tuna migrate slightly farther north. Some small fish apparently move as far east as the eastern Atlantic Ocean; that is, they mix with those spawned in the east.

From the eastern spawning ground, small fish move northward to the Bay of Biscay and southward to the Canary Islands. Some move as far west as the western Atlantic Ocean; that is, they mix with those spawned in the west.

Little is known about the movements of "medium" fish, except that they move farther eastward offshore than small fish. Practically nothing is known about movements in winter. Presumably fish move toward shore and up the coast following food or particular water temperatures. The limited evidence suggests that the distributions of small and medium fish also are patchy, with scales of patchiness being large in time and space.

---

[6]Age 0 fish are described as less than 50cm and less than 3kg, small fish are 50 to 129cm and 3 to 44kg, and medium fish are 130 to 180cm and 45 to 130kg (Clay, 1990).

Giant Bluefin Tuna in New England

Recent observations by New England commercial and recreational fishermen of large increases in local abundance may reflect large-scale changes in a food patch or substock. The New England patch may diminish within 10 years regardless of any regulations that are imposed on the fishery. The change may not represent the depletion of the stock, but rather the depletion of a food patch or other environmental changes that result in the redistribution of giant tuna.

Conclusions

There are several aspects of life history that influence the abundance indices used in management, and many aspects of life history are unknown or poorly known. There is no evidence that the large geographic separation of spawning locations represents reproductively isolated spawning grounds. Giant bluefin tuna tend to occur in patches where food is abundant and in areas of suitable temperatures. The locations of these patches appear to change from year to year.

Recommendations

NMFS, in cooperation with ICCAT, should make an effort to understand the biological and environmental requirements of spawning. Some effort should be made to estimate the relative amounts of spawning in the two locations, using identical survey methods.

Additional studies of age and size at spawning should be carried out. In particular, the committee suggests that all fish caught as by-catch in the Gulf of Mexico be sampled: the gonads should be taken for histological analysis, and body length should be measured to estimate age.

Because bluefin tuna are part of an ecosystem, the committee recommends the use of the ecosystem approach to management. For example, some effort should be directed toward estimating major changes in distribution and abundance of prey species, as prey abundance seems likely to alter the distribution and abundance of bluefin tuna. This recommendation applies to all the life history stages of bluefin tuna. This information would not necessarily be used as an index but could be useful in choosing or weighting indices or in designing various sampling schemes.

A tagging program should be undertaken by NMFS, in cooperation with ICCAT to provide better estimates of the magnitude and patterns of movement (refer to the set of design features discussed in Appendix D). This program should be designed to answer specific scientific questions pertinent to stock assessment. The program should be coordinated among *all* nations participating in ICCAT studies. Tagging should include appropriate combinations of conventional, PIT, acoustic, and archival tags (see Appendix D for a description of archival tags). Archival tags will provide information regarding fidelity to particular spawning locations and/or particular feeding locations (i.e., fishing grounds). A well-designed program using conventional tags will provide better information regarding mixing between the eastern and western Atlantic Ocean as well as better information regarding the extent that mixing changes with fish age and how it varies from year to year. It has been suggested that an aerial survey could be developed and used as a fisheries-independent estimate of abundance. Acoustic tags may be useful in calibrating aerial survey data by providing information about the vertical distribution of bluefin tuna. PIT tags (Passively Integrated Transponder) or coded wire tags may be useful in a forensic sense to estimate the magnitude of nonreporting of tag recaptures. The committee emphasizes the importance of planning tagging experiments. Much of the available information suggests that the present tagging data resulted from opportunistic programs. The committee suggests a pooling of resources among participating nations to effect a strong tagging program.

Changes in tuna distribution over time could be explained, at least in part, by large-scale changes (in time and space) in environmental conditions. These changes are discussed in the next section.

CLIMATE

Climate and Evolutionary History

One assumption underlying genetic methods used to resolve the population structure of bluefin tuna is that any of the populations are in equilibrium with migration and genetic drift (random processes). However, equilibrium is achieved, on average, only after N generations (Kimura, 1955), which is on the order of millions of years for marine fish. This assumption of equilibrium is probably not completely met for North Atlantic pelagic fish, in large part because the present-day population structure in the North Atlantic Ocean has arisen since the last glaciation (15,000 to 20,000 years ago), when an ice sheet covered the North Atlantic Ocean as far south as New England in the western Atlantic Ocean and the British Isles in the eastern Atlantic Ocean (CLIMAP, 1976). At that time, temperatures in Caribbean waters were about 5°C cooler than at present (Guilderson et al., 1994), and the present spawning areas of bluefin tuna most likely were displaced geographically.

Gene flow and the past glacial history of the North Atlantic Ocean may have been important in homogenizing populations of bluefin tuna in the North Atlantic Ocean through population contractions and recolonizations. As populations of marine fish decline, their geographic ranges contract (MacCall 1988), and gene flow over the smaller area homogenizes any genetic differences. When oceanographic conditions become more favorable, genetically homogeneous stocks recolonize the basin. Genetic differentiation has not been observed in

Atlantic herring (*Clupea harengus* [Grant, 1984]) nor in Atlantic cod (*Gadus morhua* [Grant and Stahl, 1988]), even though it is unlikely that transocean movement presently occurs in these fish. Thus, the lack of allozyme differentiation in Atlantic bluefin tuna may reflect the recent glacial history of the North Atlantic Ocean rather than ongoing migration across the Atlantic Ocean. Even if no migration occurred between western and eastern populations, allozyme frequency differences may not appear for several thousand generations. Mitochondrial and microsatellite DNA analyses, which are theoretically capable of detecting population structure on a finer scale than allozyme or mtDNA analyses, may be required to resolve the component of genetic population structure of bluefin tuna resulting from recent population events.

Conclusion

Past climatic events can have long-lasting effects on the population genetic structures of present-day populations. Molecular techniques used in studies of population structure must be capable of distinguishing historical versus contemporary factors.

Fish Abundance and Climatic Changes

Climatic variability on a shorter time scale can also influence bluefin tuna populations, but separating the effects of fishing from those of climatic change continues to challenge fisheries scientists. Although the regional abundances of some fish appear to be linked to oceanographic changes, underlying mechanisms controlling abundance are difficult to understand with data from studies that are geographically limited or of short duration. In the Pacific Ocean, fisheries fluctuations, in association with El Niño Southern Oscillations, have long been recognized. Abundances of anchoveta, and associated higher trophic levels, periodically increase in response to nutrient enrichment associated with upwelling (Barber and Smith, 1981). These upwelling events in turn, are apparently responding to pressure-driven oscillations, on the order of three to seven years, in the sea level of the western equatorial Pacific (Wyrtki, 1977).

On an interdecadal scale, the abundances of sardines and anchovies in many areas are cyclic, and are often inversely related to each other where they co-occur either in recent times (Parrish et al., 1983) or historically (Shackleton, 1987). Historical fishing records of the Japanese sardine fishery dating to the 1600s show marked fluctuations in sardine abundance (Cushing and Dickson, 1976). Strong fluctuations in abundance over the past 2,000 years have also been inferred from variability in sardine scale abundance in anaerobic sediments off California (Soutar and Isaacs, 1974; Baumgartner et al., 1992), Peru (De Vries and Pearcy, 1982), and southern Africa (Shackleton, 1987). These fluctuations in abundance predate intensive fishing on these species.

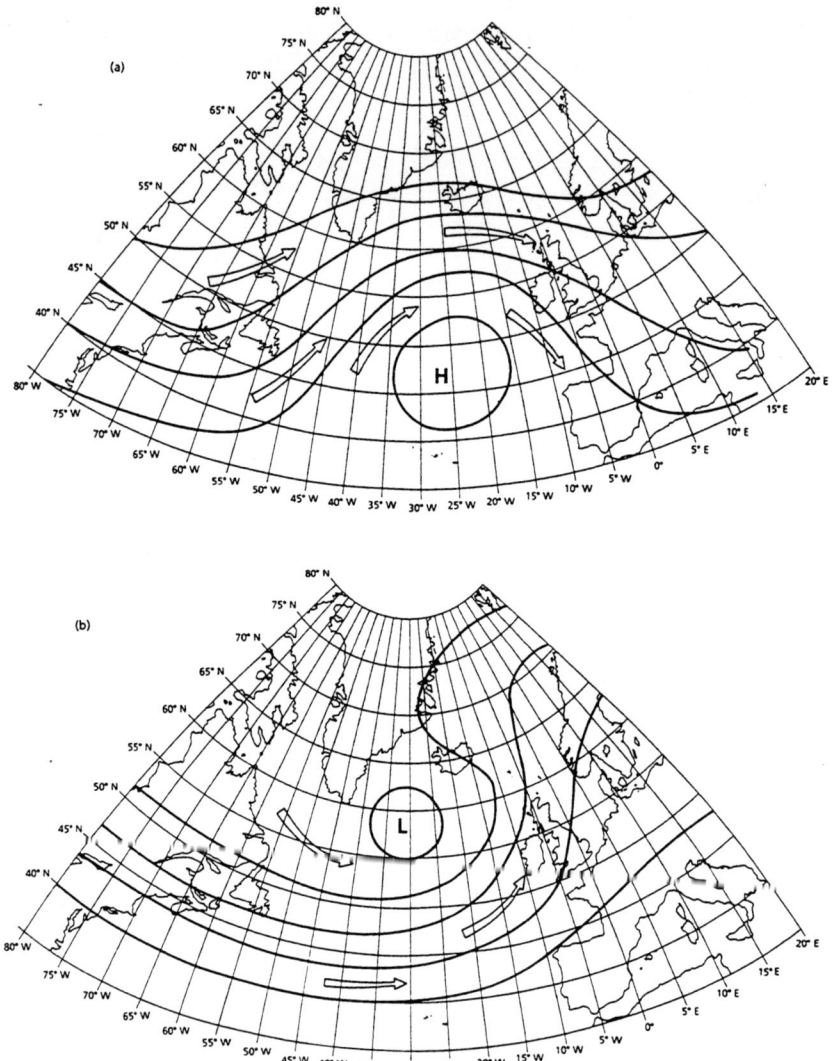

FIGURE 2-4 Barometric shifts and wind patterns driving the Russell Cycle (from Mann and Lazier [1991], with permission of Blackwell Scientific Publications Ltd., Oxford, England).

Another example of the effect of interdecadal climatic change on marine populations is the Russell Cycle in the North Atlantic. Russell observed marked changes in the abundances of macrozooplankton in the English Channel between 1924 and 1971, and suggested that local changes in abundance resulted from distributional changes associated with recurring climatically-driven oceanographic changes in the North Atlantic Ocean (Cushing and Dickson, 1976). The North Atlantic Ocean is alternately dominated by high- and low-pressure systems with a periodicity of 10-20 years. When a high-pressure system dominates, the prevailing westerly winds are diverted north of the high, resulting in a northeastward flowing body of warm air and water along North America in the western Atlantic Ocean and a southeastward flowing body of cool air and water along Europe and the British Isles (Figure 2-4). When a low-pressure system dominates in the north Atlantic Ocean, the westerlies are diverted to the south, and this results in warmer weather and sea temperature in the eastern Atlantic Ocean (Figure 2-4).

A low pressure system dominated the North Atlantic Ocean from about 1880-1935 and led to a period of gradual warming in Europe, culminating in the dramatic decade of 1925 to 1935, with the appearances of numerous tropical species along European shores and temperate species in Scandinavian waters. At this time, fishing for coldwater gadid fish was adversely influenced, and the abundance of bluefin tuna increased near Iceland (Cushing and Dickson, 1976) and in Norwegian waters (Tiews, 1978).

During this period, geographic ranges of many intertidal invertebrates were displaced to the north in the British Isles (Southward, 1980). This warm phase ended about 1935 and was followed by about 15 years of average or variable conditions, after which the North Atlantic Ocean was dominated by a high pressure system from about 1950-1970. Cooler oceanographic conditions persisted in the eastern Atlantic Ocean and gadid abundances there increased sharply despite heavy fishing. Intertidal invertebrate distributions moved southward, with coldwater boreal species replacing warmwater southern species in the British Isles (Southward, 1980). Since 1970, a low-pressure system has dominated the North Atlantic Ocean. British intertidal invertebrate ranges shifted back to the north and the fishery for gadid fishes has collapsed.

The oceanographic conditions in the western North Atlantic Ocean are more complex than those in the eastern North Atlantic Ocean. Periods of cold weather in the eastern North Atlantic Ocean are often associated with periods of warm weather in the western North Atlantic Ocean, but periods of warm eastern Atlantic Ocean weather are not always associated with cold waters in the western North Atlantic Ocean. Even though fish abundances in the western North Atlantic Ocean are not strongly correlated with climatic cycles, anecdotal reports of geographic changes in the distributions of species still abound. Tilefish, bluefish, and menhaden, which are harvested chiefly off the southeastern United States, have periodically been abundant off New England (Cushing, 1982). Hopkins and Garfield (1979), in a long-term study of temperature trends in the Gulf of Maine, suggested cycles of slightly longer than 20 years, with temperature and salinity minima in the early 1940s and the mid 1960s. This region of the western North Atlantic Ocean may be strongly influenced by the cold, southward flowing Labrador Current and by warm core rings of the northeastward-flowing Gulf Stream. Eddies of warm Gulf-Stream water can vary in number and duration before being absorbed back into the Gulf Stream (Richardson, 1983).

Although tuna are highly migratory, they are affected by climatic and oceanographic changes. Yellowfin, bigeye, and albacore, for example, show stronger year classes when sea surface temperatures are cool (Yamanaka and Yamanaka, 1970, cited in Cushing, 1982). The thermal boundaries of adult yellowfin tuna occur between 18° and 31°C, but commercial concentrations occur between 20° and 28°C (Cole, 1980). Smaller individuals of Atlantic bluefin tuna also appear to be restricted to areas in the Gulf of Maine with surface temperatures of at least 16°C, whereas adults are found in waters as cold as 10-12°C, which appears to be their lower thermal limit in the Gulf of Maine (Bigelow and Schroeder, 1953). Similarly, 12°C also appears to be the lower level of thermal tolerance for bluefin tuna in the eastern North Atlantic Ocean (Luhmann,1959; Tiews, 1962). Giant fish however, are found at temperatures as low as 6°C off Newfoundland (Squire, 1962).

Changes in oceanic temperature distributions have apparently influenced the local abundances and distributions of bluefin tuna. Bluefin tuna were first seen in Norwegian waters about 1907 at the beginning of a warming trend. This fishery peaked in 1952, with the beginning of a cooling trend, and collapsed in 1963. The collapse was attributed to overfishing (Tiews, 1978), but may have been, at least in part, owing to long-term climatic change. Farther to the south, the abundances of bluefin tuna peaked in Portuguese waters in about 1880, then decreased to one tenth this size by 1920 (Neupath, 1925). Abundance peaked again in 1937, and declined from 1958 to 1963. Although these cycles in abundance are not as strongly correlated with the Russell Cycle as are the Norwegian abundances, the declining trends may not be associated with overfishing.

These results suggest two conclusions. First, changes in abundance and geographic distribution are influenced by both climate and fishing. When abundances decline, the geographic distribution of bluefin tuna may shrink to the center of its distributional range, so that local abundance may not decline in some areas even though the overall stock is declining.

Second, even though the distributions of bluefin tuna respond to changes in sea surface temperatures, attempting to attribute changes in abundance solely to interdecadal changes in temperature may be an oversimplification. Cool water temperatures may also result from changes in upwelling, which also brings nutrient rich water to the surface and may enhance the productivity of plankton, which, in turn, may influence tuna larval abundance and year-class strengths. Climatic changes may also increase terrestrial precipitation and associated runoff with micronutrients, such as iron, that limit primary production (Gran, 1931; Martin, 1992). The important point is that climatic changes occur over wide areas and periods of many years, and these changes can affect the distributions and abundances of fish species. Fisheries managers must recognize that climatic changes can modify, for better or worse, the adverse influence of heavy fishing.

Conclusion

Changes in distribution and abundance and local declines in abundance can be brought about by changes in ocean climate, which may confound the effects of exploitation.

Recommendation

Adaptive management strategies that account for the effects of both climatic changes and exploitation should be considered for implementation. Such strategies should also account for the dynamic nature of geographic distributions in space and time.

MOVEMENT BASED ON NONGENETIC MARKERS

Nongenetic markers can provide only indirect evidence for gene flow across the Atlantic Ocean because there is no assurance that fish moving from one area to the next actually breed in the new area; mixing on the fishing grounds may not necessarily reflect mixing of the same magnitude, or mixing of the same fish, on the spawning grounds. Such data can, nonetheless, provide circumstantial evidence for potential interbreeding among stocks. Naturally occurring markers of population mixing include parasites and chemical constituents, both of which must be acquired at an early age in nursery areas to be useful for discrimination among stocks. Tag and recapture experiments are specific attempts to estimate movement from one area to another and can be used at any age of the fish. Such data constitute conclusive evidence of movement from the area of tagging to the area of recapture.

Parasite Markers

The geographic distributions of parasites on marine fish have been used to infer the evolution and migratory patterns of the fish (e.g., Kabata and Ho, 1981). For this approach to be useful in determining the geographic origin of a fish, the fish must acquire a parasite at an early age and the parasite must have a restricted geographic distribution where infection is possible. The trematode *Nasicola sp.* infests the nasal pores of the bluefin tuna and is tropical, so that infection is possible only in the Gulf of Mexico or in other tropical waters of the Atlantic Ocean. The copepod *Elytrophora sp.* infests the gill chamber of bluefin tuna, lives in temperate waters, and cannot be acquired in the tropics. Walters (1980) conducted a study of these parasites in Atlantic bluefin tuna and the results of his work are often cited as evidence of movement of bluefin tuna between areas in the north Atlantic Ocean. A close scrutiny of these data, however, show that his results cannot be used to infer movement (Appendix E).

Conclusion

Existing data on parasite markers do not provide evidence to support the mixing hypothesis.

Recommendation

This avenue of research is not recommended until experts in the field can show its utility and efficacy for estimating movement of bluefin tuna.

Microconstituent Analysis

The idea behind this method of estimating population structure is that the proportions of chemical elements incorporated into fish bones differ from one area to the next because of differences in elemental compositions of prey and seawater. Thus, each nursery area may have a different elemental signature in bony material laid down during growth, and the signature can be used to identify the origins of fish captured outside the nursery areas. Calaprice (1985) and Calaprice et al. (1971) measured elemental signatures by x-ray fluorospectroscopy. Energy from gamma rays is absorbed by electrons in the various elements embedded in bone and is spontaneously released to produce characteristic x-ray signatures at lower energies. The x-ray emission spectrum is recorded and used to characterize individual fish.

The two Calaprice studies are concordant with the hypothesis of significant movement between east and west. However, they cannot be used to assign a value to the magnitude of the mixing because the results vary markedly between studies. The reason for the variability of the studies probably resides in the nonrandom procedures used to obtain fish samples from the different fishing grounds. The rationale for the committee's position is detailed in Appendix F.

Conclusions

The microconstituent studies support the hypothesis of mixing between the two regions. Some giant fish harvested in the western Atlantic Ocean were spawned in the Mediterranean Sea, and some that were harvested in the eastern Atlantic Ocean were spawned in the Gulf of Mexico.

The data suggest that the birthplace of older tuna can be estimated by analyzing the center of the vertebrae because some aspects of the chemical "fingerprint" are stable over time.

The available data and analysis cannot be used to ascribe a value to the magnitude of mixing in either direction.

Recommendation

This avenue of research should be pursued by NMFS and ICCAT because the preliminary results look promising. There have been marked improvements in techniques similar to those employed by Calaprice and in software used to discern patterns. Any research proposal regarding this complex analytical technique should be peer reviewed. Calaprice noted that "the acquisition of adequate samples has been a difficult and limiting task." Any future research should engage all nations so that adequate samples from all areas can be used in the analysis. A single-blind approach should be used so that the persons doing the analysis do not know the source of the material but are aware of the number of localities that have been sampled. The observation that bluefin tuna often are found in schools (or larger patches) means that each sample must cut across schools and patches. A random sample from the western Atlantic Ocean must include samples from several localities, and not just from a single school.

3

Transatlantic Movement of Atlantic Bluefin Tuna

INTRODUCTION

The degree of fish movement between fishing grounds in the eastern and western Atlantic Ocean and the degree of genetic mixing in the spawning areas in the eastern and western Atlantic Ocean are both important topics relevant to the management of Atlantic bluefin tuna. The review and reanalysis of tagging data constitute the committee's evaluation of the scientific basis for physical movement or mixing of fish between fishing grounds. These tagging data, however, do not provide information on the degree of genetic mixing.

TAG-RECAPTURE DATA

Tag-recapture data provide the strongest evidence available for transatlantic movement of fish. Atlantic bluefin tuna were marked and released in the western Atlantic Ocean from 1954 in varying numbers until 1987 (Parrack, 1990). The total number tagged by the National Marine Fisheries Service (NMFS) since the inception of the program has been over 15,000. Early releases were primarily of large (giant) fish, although more recently releases have included numerous small fish.

West to East

An overview of movement from the west to the east by size class (i.e., from age 0 fish to giant fish), as indicated by tag returns, appears in Suzuki (1990).

TABLE 3-1 Synopsis of release and recapture (tagging experiments) of western Atlantic bluefin tuna. Data are from Mather (1980) and Brunenmeister (1980).

Tag locality	# Released	Size	# Recaptured	
Bahamas	1,709	Giant fish	17	
NW Atlantic Ocean	1,881	Giant fish	81	
US Coast*	468	Medium fish	10	
NW Atlantic Ocean	>17,700	Small fish	>2,180	
Bahamas to eastern Atlantic Ocean		9	[giant fish]	
US Coast* to eastern Atlantic Ocean		1	[medium fish]	
NW Atlantic to eastern Atlantic Ocean		46	[small fish]	
Trans-Atlantic giant fish released in Bahamas			9/17	52.9%
Trans-Atlantic giant fish released in W Atlantic Ocean			9/98	9.2%
Trans-Atlantic medium fish released off US Coast*			1/10	10.0%
Trans-Atlantic small fish released in NW Atlantic Ocean			46/2,180	2.1%**

*From Mather (1980): "mostly" medium fish were released north of 35°N and west of 60°W.
**More than 2,180 fish released; estimate of 2.1% is maximum.

Recaptures through 1978 of large fish released in the Bahamas in the 1960s, and of small fish released along the U.S. Atlantic coast, document unequivocally that transatlantic migration occurs (Table 3-1). Transatlantic migrants include nine giant fish taken off the Norwegian coast, one medium fish (recovered as a giant tuna over 10 years later) taken off the southern coast of Spain, and 46 small fish taken chiefly from the Bay of Biscay. Estimates of the total proportion of fish tagged in the western Atlantic Ocean in eastern Atlantic samples (Table 3-1) range from 9.3% for giant fish and medium fish to 2.1% for small fish. Estimates for giant and medium fish differ slightly from those given in Mather (1980), who found a total of 10 western migrants in 126 eastern Atlantic recaptures (10/126 = 7.9%). Suzuki (1990) estimated an "overall" proportion of western-tagged fish in eastern Atlantic samples of 3.2% as of 1988.

Estimates of west to east transatlantic movement are variable and depend on the size of fish tagged and tagging locality. All but one of the giant fish transatlantic recoveries were tagged near the Bahamas, and five of the nine migrants were recovered in Norwegian waters in the same year. These results indicate that giant bluefin tuna can travel long distances; only one of the remaining 91 giant tuna or medium fish tagged in the western Atlantic Ocean was recovered in the eastern Atlantic Ocean. This fish was tagged near Bermuda and recovered 10 years later. Mather (1980) hypothesized that the transatlantic migration of giant fish was not part of an annual pattern but rather represented irregular migration of varying numbers of fish in different years. This hypothesis deserves consideration because bluefin tuna tend to follow ocean currents (Sella, 1929, cited in Mather, 1980) and because of the proximity of the Bahamian tagging sites to the Gulf Stream. This possible transatlantic "route" for giant fish is strikingly similar to the one for larvae proposed by Murphy (1990). Finally, Mather (1980) also noted that the transatlantic movement of small bluefin tuna was variable and that virtually all transatlantic movements of small fish were from the Mid-Atlantic Bight to the Bay of Biscay. Mather concluded that the west to east transatlantic movement of small fish might be due to unusual stimuli and that the great majority of Atlantic bluefin tuna remain on one side of the Atlantic Ocean or the other. Alternatively, the "nonrandom" movement of small bluefin tuna from the Mid-Atlantic Bight to the Bay of Biscay may reflect the importance of both nursery areas for small bluefin tuna.

East to West

Eastern Atlantic/Mediterranean bluefin tuna have been tagged and released since 1911, although most of the tag and release data are from 1957 to the present (Cort and Liorzou, 1990c). As in the western Atlantic Ocean, most of the early releases were giant tuna, with the number of small fish released and recovered increasing over the past two decades. Numbers of fish released and recaptured from the eastern Atlantic Ocean and Mediterranean Sea from 1911 through 1992 are listed in Brunenmeister (1980), Cort and Liorzou (1990c), and Cort and de la Serna (1993). A synopsis of these listings is shown in Table 3-2. The proportion of eastern-tagged fish among western Atlantic recaptures varies from 0% for giant fish to 4.5% for small fish. Virtually all of the transatlantic migrants recaptured off the U.S. coast were released from the Bay of Biscay, and most were captured between 10 and 20 months after first released. The "total" proportion of eastern fish among western Atlantic recaptures is 4.0% (19/472) after adjustment for the greater number of small fish among recaptures. This estimate is similar to the estimate of 4.4% of Cort and Loirzou (1990c). Estimates of eastern Atlantic fish among Mediterranean recaptures are 11.3% for giant fish and 2.8% for small fish. The proportion of Mediterranean fish among eastern Atlantic recaptures is 2.8% for small fish.

The absence of documented east to west transatlantic movement of large fish is striking and merits further investigation. Considering only small fish, the proportion of transatlantic immigrants in western and eastern Atlantic samples differs: estimates made here suggest an overall proportion of small western fish among eastern Atlantic recaptures of 2.1% and a proportion of 4.5% eastern-tagged fish among western recaptures.

TABLE 3-2 Synopsis of release and recapture (tagging experiments) of eastern Atlantic/Mediterranean bluefin tuna. Data are from Brunenmeister (1980), Cort and Loirzou (1990c), and Cort and de la Serna (1993).

Tag Locality	# Released	Size	# Recaptured	
Eastern Atlantic Ocean	599-604	Giant fish	53	
	6,144	Small fish	418-420	
	107-232	Unknown*	3	
Mediterranean Sea	3,993	Small fish	70	
	20-30	Unknown*	0	
Eastern Atlantic to western Atlantic Ocean			19	[small fish]
Eastern Atlantic Ocean to Mediterranean Sea			13	[6 giant fish, 7 small fish]
Mediterranean Sea to eastern Atlantic Ocean			11	[small fish]
Trans-Atlantic giant fish released in eastern Atlantic Ocean			0	
Trans-Atlantic small fish relesed in eastern Atlantic Ocean			19/419	4.5%
Mediterranean small fish to eastern Atlantic Ocean			11/70	15.7%
Eastern Atlantic giant fish to Mediterranean Sea			6/53	11.3%
Eastern Atlantic small fish to Mediterranean Sea			13/472	2.8%

* From studies carried out prior to 1940 in Italy and Portugal.

Reanalysis of Tagging Data

Tagging data suggest that physical movement or mixing of Atlantic bluefin tuna in fishing grounds from opposite sides of the Atlantic Ocean is significant. Thus, the committee reanalyzed the tagging data to estimate rates of transatlantic movement between fishing grounds. This analysis was undertaken to provide rigorous estimates of transfer across the Atlantic Ocean, which has not been attmepted previously. Because it is not possible to estimate transfer rates without knowledge of the total mortality rates and the nonreporting rates (tags recovered but not reported), these quantities are also estimated.

METHODS

The problem of transfer rates is described by the rates of change in the population sizes, N_w for the west and N_e for the east. Equations for the transfer of fish from the west to the east, applicable to fish tagged with a single tag, are:

$$dN_w/dt = -(M + F_w + T_w)N_w \tag{1}$$

$$dN_e/dt = -(M + F_e)N_e + T_w N_w \qquad (2)$$

M = annual rate of instantaneous natural mortality plus shedding.[1]
F_w = annual rate of instantaneous fishing mortality in the west.
F_e = annual rate of instantaneous fishing mortality in the east.
T_w = annual rate of instantaneous transfer from west to east.

Integration of this pair of equations provides estimates of the population size at any time t, over the interval for which F_w is constant in the west and F_e is constant in the east. The solutions are given by:

$$N_w(t) = N_w(0)\exp(-Zt), \text{ where } Z = M + F_w + T_w$$

$$N_e(t) = T_w N_w(0)\exp(-Xt)[1 - \exp(-Yt)]/Y,$$

where $X = M + F_e$ and $Y = F_w - F_e + T_w$

Similarly, the equations for the catches are:

$$C_w(t) = N_w(0)F_w[1 - \exp(-Zt)]/Z$$

$$C_e(t) = N_w(0)F_e T_w \{[1 - \exp(-Xt)]/X - [1 - \exp(-Zt)]/Z\}/Y$$

When t is sufficiently large:

$$C_e/C_w = T_w F_e/[F_w(F_e + M)]$$

When P_w and P_e are the proportions of tags recovered that are reported for the west and east, respectively, then:

$$C_e/C_w = P_e T_w F_e/[P_w F_w(F_e + M)] \qquad (3)$$

Analogous equations (with the subscripts e and w reversed) are used to estimate the transfer rate of fish from the east to the west. Because the models given by Equations (1) and (2) do not admit roundtrip movement, any such return movement would cause the transfer rates estimated by Equation (3) to be underestimates of movement rates from one side of the ocean to the other. Also, if fishing mortality decreases with age, as it appears to do, the movement rates estimated by assuming a constant fishing mortality will be underestimated.

Tag-recovery information from Corte and de la Serna (1993) (Table 2; which is reproduced below as Table 3-3) and from NMFS archives (file name MRFISH)

---

[1] Shedding refers to tags that have detached from the fish.

TABLE 3-3 Spanish tagging data for Atlantic bluefin tuna in the Cantabrian Sea (Bay of Biscay) from 1976 to 1991.

Year	N	Recoveries in the Eastern Atlantic With Known Years At Liberty					Trans Atlantic	Med. Sea	East Atlantic*	No data	Total
		0	1	2	3	4					
1976	3	-	-	-	-	-	0	-	0	0	-
1977	10	-	-	-	-	-	0	-	0	0	-
1978	170	29	2	4	1	-	1	-	0	5	42
1979	101	1	10	-	-	-	2	-	0	1	14
1980	302	15	2	2	-	-	3	-	2	8	32
1981	293	3	5	-	1	-	0	2	0	2	13
1982	395	5	6	4	2	-	1	3	2	2	25
1983	370	2	1	1	-	-	0	1	0	1	6
1984	513	8	7	1	-	1	0	1	2	6	26
1985	407	12	2	-	2	-	1	-	0	4	21
1986	838	37	8	6	1	1	5	-	0	10	68
1987	-	-	-	-	-	-	0	-	0	0	-
1988	1,151	26	17	5	-	2	1	-	0	7	58
1989	122	2	1	2	2	-	0	-	0	0	7
1990	973	8	22	2	-	-	3	-	0	0	39
1991	15	4	1	-	-	-	0	-	0	0	5
Total	5,663	152	84	29	11	4	17	7	6	46	356

*This column refers to recoveries in the eastern Atlantic without known recovery year.

was used to obtain values of catch (C). Some descriptions of qualitative features of the tagging data with respect to movement are given below in the Results section. Shedding ($L = M - 0.14$) (0.14 was taken from the literature as the natural mortality rate; ICCAT, 1993) was estimated from these catch data using the method of Chapman, Fink, and Bennet (1965, equations 5, 6, and 7) to estimate L. This method was used rather than that of Bayliff and Mobrand (1972), because shedding rates do not appear to be constant and appear to be insensitive to type I (initial) shedding. Once the estimate of M ($L + 0.14$) was available, VPA analysis was done on the catches over time to obtain estimates of F.

Results

Tables 3-4, 3-5, and 3-6 show the number of fish tagged in the western Atlantic Ocean (from MRFISH) and the number of recoveries by year after tagging. Year 0 means fish caught in the same calendar year as tagged; year 1 means fish caught in the calendar year following tagging, etc. Table 3-7 (eastern Atlantic Ocean, ICCAT data) and Table 3-8 (western Atlantic Ocean, NMFS data) show the months in which the fish were tagged and the months in which the fisheries captured the tagged fish. Because most tagged fish were small (less than four years old), inferences about fishing times based on these tag recoveries should apply to small fish.

From 1971 to 1978, there was intensive tagging in the west with both single- and double-tagged fish (a small number of double-tagged fish actually had more than two tags). Table 3-9 shows the number of recoveries from single-tagged fish (SS), the number of recoveries of fish with two tags from double-tagged fish (DD), and the number of recoveries of fish with one tag from double-tagged fish (DS), by year of recovery. Tagging time varied, so data from the west were sorted into periods of one-quarter year; that is, if a tag was recovered with the same quarter as the release, it was placed in quarter 1. If the recovery was made after more than 91 days but before 182 days, it was placed in quarter 2, and so on. Table 3-10 shows the recovery data for recoveries SS, DD, and DS by year of release and quarter of recovery.

The western data from Table 3-10 were used to estimate shedding rates (Table 3-11). Comparison of SS to DD (Db Db vs. Sing in Table 3-11) shows the most variability, and comparison of DD to DS (Db Db vs. Db Sing) shows the most stability. After examining Table 3-11, it was decided to try $L = 0.26$ for all ages and also to try $L(1) = 0.50$, $L(2) = 0.40$, $L(3) = 0.30$, and $L(4+) = 0.20$, corresponding to quarter 1, quarter 2, quarter 3, and all quarters 4 or more. These values are not much different from the value of 0.205 estimated by Baglin et al. (1980).

With the shedding rates mentioned above, and the catch vectors for SS and DD in Table 3-11, VPA analysis was performed showing that the two sets of

TABLE 3-4 Atlantic bluefin tuna release and recovery data from the United States tagging program in the western Atlantic Ocean. Tag returns are by year tagged and years out for all tag types.

Year	Tag	0	1	2	3	4	5	6	7	8
1954	193	1	-	-	-	-	2	-	-	-
1955	232	-	-	-	-	-	-	-	-	-
1956	99	-	-	-	-	-	-	-	-	-
1957	39	-	-	1	-	-	-	-	-	-
1958	38	-	-	-	-	-	-	-	-	-
1959	147	1	-	-	-	-	-	-	-	-
1960	237	-	-	3	1	-	-	-	-	1
1961	188	2	2	3	2	-	-	-	-	-
1962	128	1	4	-	-	-	-	-	-	-
1963	223	11	3	1	-	-	-	-	-	-
1964	553	100	32	-	-	-	-	-	-	-
1965	1,812	164	63	36	3	-	-	-	-	-
1966	4,128	524	577	51	8	6	-	2	-	-
1967	718	98	60	16	13	-	-	-	-	-
1968	521	88	19	9	-	-	-	-	-	-
1969	567	14	81	12	2	2	-	1	1	-
1970	729	52	119	9	7	3	-	-	-	-
1971	432	11	59	12	-	-	1	-	-	-
1972	284	10	56	6	2	-	-	-	-	-
1973	393	40	22	7	2	-	-	-	-	-
1974	1,752	75	126	51	14	7	-	-	-	-
1975	349	22	27	6	2	-	-	1	-	-
1976	2,460	185	63	75	8	-	1	-	-	1
1977	2,115	52	193	81	2	-	-	1	2	1
1978	1,680	24	110	34	1	7	3	1	1	?
1979	1,124	6	30	11	1	3	1	1	-	1
1980	3,074	141	100	6	3	4	2	2	4	-
1981	1,797	57	2	2	1	1	1	2	-	4
1982	210	1	1	1	2	-	-	-	-	2
1983	149	1	2	4	4	-	-	1	-	-
1984	86	1	2	-	-	-	-	-	-	-
1985	130	-	1	2	-	-	-	-	-	-
1986	51	-	1	1	-	-	-	-	-	-
1987	66	1	1	-	-	-	-	-	-	-
1988	96	-	2	-	-	-	-	-	-	-
1989	113	-	-	-	-	-	-	-	-	-
1990	127	-	-	-	-	-	-	-	-	-

Note: Dash indicates zero value.

9	10	11	12	13	14	15	16	17	18	Total
-	-	-	-	-	-	-	-	-	-	3
-	-	-	-	-	-	-	-	-	-	-
-	-	-	-	-	-	-	-	-	-	-
-	-	-	-	-	-	-	-	-	-	1
-	-	-	-	-	-	-	-	-	-	-
-	-	-	-	-	-	-	-	-	-	1
-	-	-	-	-	-	-	-	-	-	5
-	-	-	-	-	-	-	-	-	-	9
-	-	-	-	-	-	-	-	-	-	5
-	-	-	-	-	-	-	-	-	-	15
-	-	-	-	-	-	-	-	-	-	132
-	-	-	-	-	-	-	-	-	-	267
1	1	-	-	2	-	-	-	-	-	1,172
-	-	-	-	-	-	-	-	-	1	188
-	-	-	-	-	-	-	-	-	-	116
1	-	-	-	-	-	-	-	-	-	114
-	-	-	-	-	-	-	-	-	-	190
-	-	-	-	-	-	-	-	-	-	83
-	-	-	-	-	-	-	-	-	-	74
-	-	-	-	-	-	-	-	-	-	71
1	1	1	1	-	-	1	-	-	-	278
-	-	-	-	-	-	-	-	-	-	58
-	-	-	-	-	-	-	-	-	-	333
1	1	2	-	-	-	-	-	-	-	341
-	1	2	-	-	-	-	-	-	-	192
-	-	-	-	-	-	-	-	-	-	53
2	-	-	-	-	-	-	-	-	-	268
-	-	-	-	-	-	-	-	-	-	68
-	-	-	-	-	-	-	-	-	-	5
-	-	-	-	-	-	-	-	-	-	12
-	-	-	-	-	-	-	-	-	-	3
-	-	-	-	-	-	-	-	-	-	3
-	-	-	-	-	-	-	-	-	-	2
-	-	-	-	-	-	-	-	-	-	2
-	-	-	-	-	-	-	-	-	-	2
-	-	-	-	-	-	-	-	-	-	-
-	-	-	-	-	-	-	-	-	-	-

TABLE 3-5 Atlantic bluefin tuna release and recovery data from the U.S. tagging program in the western Atlantic Ocean. Tag returns are by year tagged and years out for fish tagged with a single tag.

Year	Tag	0	1	2	3	4	5	6	7	8
1954	192	1	-	-	-	-	2	-	-	-
1955	231	-	-	-	-	-	-	-	-	-
1956	99	-	-	-	-	-	-	-	-	-
1957	39	-	-	1	-	-	-	-	-	-
1958	38	-	-	-	-	-	-	-	-	-
1959	147	1	-	-	-	-	-	-	-	-
1960	237	-	-	3	1	-	-	-	-	1
1961	185	2	2	3	2	-	-	-	-	-
1962	128	1	4	-	-	-	-	-	-	-
1963	183	7	3	1	-	-	-	-	-	-
1964	544	100	32	-	-	-	-	-	-	-
1965	1,751	164	63	36	3	-	-	-	-	-
1966	4,114	523	575	51	8	6	-	2	-	-
1967	718	98	60	16	13	-	-	-	-	-
1968	448	60	12	6	-	-	-	-	-	-
1969	547	14	77	12	1	2	-	1	1	-
1970	724	51	119	9	7	2	-	-	-	-
1971	115	2	11	3	-	-	1	-	-	-
1972	130	3	14	6	1	-	-	-	-	-
1973	105	4	8	1	1	-	-	-	-	-
1974	466	17	43	14	7	4	-	-	-	-
1975	239	20	16	4	-	-	-	1	-	-
1976	1,916	145	52	66	5	-	1	-	-	-
1977	1,590	39	129	50	1	-	-	-	1	1
1978	848	20	36	13	4	4	1	1	-	-
1979	1,110	5	29	10	1	3	1	1	-	-
1980	3,066	141	100	6	3	4	2	2	2	4
1981	1,787	55	2	2	1	1	1	1	-	2
1982	200	1	1	1	2	-	-	-	-	-
1983	146	1	1	4	4	-	-	1	-	-
1984	84	-	2	-	-	-	-	-	-	-
1985	128	-	1	2	-	-	-	-	-	-
1986	51	-	1	1	-	-	-	-	-	-
1987	66	1	1	-	-	-	-	-	-	-
1988	96	-	2	-	-	-	-	-	-	-
1989	112	-	-	-	-	-	-	-	-	-
1990	127	-	-	-	-	-	-	-	-	-

Note: Dash indicates zero value.

9	10	11	12	13	14	15	16	17	18	Total
-	-	-	-	-	-	-	-	-	-	3
-	-	-	-	-	-	-	-	-	-	-
-	-	-	-	-	-	-	-	-	-	-
-	-	-	-	-	-	-	-	-	-	1
-	-	-	-	-	-	-	-	-	-	-
-	-	-	-	-	-	-	-	-	-	1
-	-	-	-	-	-	-	-	-	-	5
-	-	-	-	-	-	-	-	-	-	9
-	-	-	-	-	-	-	-	-	-	5
-	-	-	-	-	-	-	-	-	-	10
-	-	-	-	-	-	-	-	-	-	132
-	-	-	-	-	-	-	-	-	-	267
1	1	-	-	2	-	-	-	-	-	1,169
-	-	-	-	-	-	-	-	-	1	188
-	-	-	-	-	-	-	-	-	-	78
1	-	-	-	-	-	-	-	-	-	109
-	-	-	-	-	-	-	-	-	-	188
-	-	-	-	-	-	-	-	-	-	17
-	-	-	-	-	-	-	-	-	-	21
-	-	-	-	-	-	-	-	-	-	14
1	-	-	1	-	-	-	-	-	-	87
-	-	-	-	-	-	-	-	-	-	41
-	-	-	-	-	-	-	-	-	-	269
1	1	1	-	-	-	-	-	-	-	224
-	-	1	-	-	-	-	-	-	-	80
-	-	-	-	-	-	-	-	-	-	50
4	2	-	-	-	-	-	-	-	-	268
-	-	-	-	-	-	-	-	-	-	65
-	-	-	-	-	-	-	-	-	-	5
-	-	-	-	-	-	-	-	-	-	11
-	-	-	-	-	-	-	-	-	-	2
-	-	-	-	-	-	-	-	-	-	3
-	-	-	-	-	-	-	-	-	-	2
-	-	-	-	-	-	-	-	-	-	2
-	-	-	-	-	-	-	-	-	-	2
-	-	-	-	-	-	-	-	-	-	-
-	-	-	-	-	-	-	-	-	-	-

TABLE 3-6 Atlantic bluefin tuna release and recovery data from the U.S. tagging program in the western Atlantic Ocean. Tag returns are by year tagged and years out for fish tagged with two or more tags.

Year	Tag	0	1	2	3	4	5	6	7	8
1954	1	-	-	-	-	-	-	-	-	-
1955	1	-	-	-	-	-	-	-	-	-
1961	3	-	-	-	-	-	-	-	-	-
1963	40	4	1	-	-	-	-	-	-	-
1964	9	-	-	-	-	-	-	-	-	-
1965	61	-	-	-	-	-	-	-	-	-
1966	14	1	2	-	-	-	-	-	-	-
1968	73	28	7	3	-	-	-	-	-	-
1969	20	-	4	-	1	-	-	-	-	-
1970	5	1	-	-	-	1	-	-	-	-
1971	317	9	48	9	-	-	-	-	-	-
1972	154	7	42	3	1	-	-	-	-	-
1973	288	36	14	6	1	-	-	-	-	-
1974	1,286	58	83	37	7	3	-	-	-	-
1975	110	2	11	2	2	-	-	-	-	-
1976	544	40	11	9	3	-	-	-	-	1
1977	525	13	69	31	1	-	-	1	1	-
1978	832	4	74	21	3	3	2	-	1	2
1979	14	1	1	1	-	-	-	-	-	-
1980	8	-	-	-	-	-	-	-	-	-
1981	10	2	-	-	-	-	-	1	-	-
1982	10	-	-	-	-	-	-	-	-	-
1983	3	-	1	-	-	-	-	-	-	-
1984	2	1	-	-	-	-	-	-	-	-
1985	2	-	-	-	-	-	-	-	-	-
1989	1	-	-	-	-	-	-	-	-	-

Note: Dash indicates zero value.

9	10	11	12	13	14	15	16	17	18	Total
-	-	-	-	-	-	-	-	-	-	-
-	-	-	-	-	-	-	-	-	-	-
-	-	-	-	-	-	-	-	-	-	-
-	-	-	-	-	-	-	-	-	-	5
-	-	-	-	-	-	-	-	-	-	-
-	-	-	-	-	-	-	-	-	-	-
-	-	-	-	-	-	-	-	-	-	3
-	-	-	-	-	-	-	-	-	-	38
-	-	-	-	-	-	-	-	-	-	5
-	-	-	-	-	-	-	-	-	-	2
-	-	-	-	-	-	-	-	-	-	66
-	-	-	-	-	-	-	-	-	-	53
-	-	-	-	-	-	-	-	-	-	57
-	1	1	-	-	-	1	-	-	-	191
-	-	-	-	-	-	-	-	-	-	17
-	-	-	-	-	-	-	-	-	-	64
-	-	1	-	-	-	-	-	-	-	117
-	1	1	-	-	-	-	-	-	-	112
-	-	-	-	-	-	-	-	-	-	3
-	-	-	-	-	-	-	-	-	-	-
-	-	-	-	-	-	-	-	-	-	3
-	-	-	-	-	-	-	-	-	-	-
-	-	-	-	-	-	-	-	-	-	1
-	-	-	-	-	-	-	-	-	-	1
-	-	-	-	-	-	-	-	-	-	-
-	-	-	-	-	-	-	-	-	-	-

TABLE 3-7 Atlantic bluefin tuna tagged in the eastern Atlantic Ocean from 1976 to 1991. Month of tagging versus month of recapture, all types of tag.

Month Tagged	Month of Recovery												Total
	1	2	3	4	5	6	7	8	9	10	11	12	
1	-	-	-	-	-	-	-	-	-	-	-	-	-
2	-	-	-	-	-	-	-	-	-	-	-	-	-
3	-	-	-	-	-	-	-	-	-	-	-	-	-
4	-	-	-	-	-	-	-	-	-	-	-	-	-
5	-	-	-	-	-	-	-	-	-	-	-	-	-
6	-	-	-	-	-	-	-	-	-	-	-	-	-
7	-	-	-	-	1	-	1	2	3	6	3	-	16
8	1	-	-	-	-	2	16	31	30	27	7	-	114
9	-	-	-	-	-	1	1	2	-	3	10	1	18
10	-	-	-	-	-	1	3	3	1	6	26	-	40
11	-	2	-	-	-	-	1	1	1	-	13	23	41
12	-	-	-	-	-	-	-	-	-	-	-	-	-
Total	1	2	-	-	1	4	22	39	35	42	59	24	229

Note: Dash indicates zero value.

TABLE 3-8 Atlantic bluefin tuna tagged in the western Atlantic Ocean from 1954 to 1990. Month of tagging versus month of recapture.

Month Tagged	Month of Recovery												Total
	1	2	3	4	5	6	7	8	9	10	11	12	
1	-	-	-	-	-	-	-	-	-	-	-	-	-
2	-	-	-	-	-	-	-	-	-	-	-	-	-
3	-	-	-	-	-	-	1	-	-	-	-	-	1
4	-	-	-	-	-	-	-	-	-	-	-	-	-
5	-	-	2	2	-	1	1	3	-	-	-	1	10
6	11	2	-	2	3	157	36	104	60	5	1	23	404
7	16	6	-	4	3	304	739	644	177	80	1	41	2,015
8	11	5	1	2	1	127	320	723	157	32	-	10	1,389
9	-	4	4	-	-	9	76	86	8	9	-	2	218
10	-	1	-	-	-	3	9	13	-	1	-	-	7
11	-	-	-	-	-	1	-	1	-	-	-	-	2
12	-	-	-	-	-	-	-	-	-	-	-	-	-
Total	38	18	7	10	7	602	1,182	1,574	422	127	2	77	4,066

Note: Dash indicates zero value.

TABLE 3-9 Atlantic bluefin tuna release and recovery data from the U.S. tagging program in the western Atlantic Ocean. Tag returns are by year tagged and years out for single tags, double tags recovered with two tags, and double tags recovered with single tags.

SINGLE TAGGED - SINGLE RECOVERY

		Recoveries by Year						
Year	Tagged	0	1	2	3	4	5	6
1971	115	2	11	3	-	-	1	-
1972	130	3	14	3	1	-	-	-
1973	105	4	8	1	1	-	-	-
1974	466	17	43	14	7	4	-	-
1975	239	20	16	4	-	-	-	1
1976	1,916	145	52	66	5	-	1	-
1977	1,590	39	129	50	1	-	-	-
1978	848	20	36	13	4	4	1	1
Total	5,409	250	309	154	19	8	3	2

DOUBLE TAGGED - DOUBLE RECOVERY

		Recoveries by Year						
Year	Tagged	0	1	2	3	4	5	6
1971	317	8	30	5	-	-	-	-
1972	154	6	27	1	-	-	-	-
1973	288	29	7	2	-	-	-	-
1974	1,286	52	57	19	3	-	-	-
1975	110	1	5	2	-	-	-	-
1976	544	36	8	3	2	-	-	-
1977	525	12	56	23	-	-	-	-
1978	832	4	61	9	1	1	-	-
Total	4,056	148	251	64	6	1	-	-

DOUBLE TAGGED - SINGLE RECOVERY

		Recoveries by Year						
Year	Tagged	0	1	2	3	4	5	6
1971	317	1	18	4	-	-	-	-
1972	154	1	15	2	1	-	-	-
1973	288	7	7	4	1	-	-	-
1974	1,286	6	26	18	4	3	-	-
1975	110	1	6	-	2	-	-	-
1976	544	4	3	6	1	-	-	-
1977	525	1	13	8	1	-	-	1
1978	832	-	13	12	2	2	2	-
Total	4,056	21	101	54	12	5	2	1

Note: Dash indicates zero value.

	7	8	9	10	11	12	13	14	15	Total
	-	-	-	-	-	-	-	-	-	17
	-	-	-	-	-	-	-	-	-	21
	-	-	-	-	-	-	-	-	-	14
	-	-	1	-	-	1	-	-	-	87
	-	-	-	-	-	-	-	-	-	41
	-	-	-	-	-	-	-	-	-	269
	1	1	1	1	1	-	-	-	-	224
	-	-	-	-	1	-	-	-	-	80
	1	1	2	1	2	1	-	-	-	753

	7	8	9	10	11	12	13	14	15	Total
	-	-	-	-	-	-	-	-	-	43
	-	-	-	-	-	-	-	-	-	34
	-	-	-	-	-	-	-	-	-	38
	-	-	-	-	1	-	-	-	1	133
	-	-	-	-	-	-	-	-	-	8
	-	-	-	-	-	-	-	-	-	49
	1	-	-	-	-	-	-	-	-	92
	1	-	-	-	-	-	-	-	-	77
	2	-	-	-	1	-	-	-	1	474

	7	8	9	10	11	12	13	14	15	Total
	-	-	-	-	-	-	-	-	-	23
	-	-	-	-	-	-	-	-	-	19
	-	-	-	-	-	-	-	-	-	19
	-	-	-	1	-	-	-	-	-	58
	-	-	-	-	-	-	-	-	-	9
	-	1	-	-	-	-	-	-	-	15
	-	-	-	-	1	-	-	-	-	25
	-	2	-	1	1	-	-	-	-	35
	-	3	-	2	2	-	-	-	-	203

TABLE 3-10 Atlantic bluefin tuna release and recovery data from the U.S. tagging program in the western Atlantic Ocean. Tag returns are by year tagged and quarters out for fish tagged with a single tag, tagged with two tags and recovered with two tags, and tagged with two tags and returned with one tag.

SINGLE TAGGED - SINGLE RECOVERY

QR	Year Tagged								
	1971	1972	1973	1974	1975	1976	1977	1978	TOTAL
1	2	2	4	15	15	138	35	17	228
2	-	1	-	4	5	5	4	3	22
3	-	-	-	-	-	4	7	1	12
4	1	-	2	15	3	8	49	32	110
5	10	11	5	26	13	37	65	3	170
6	-	2	1	1	-	1	7	-	12
7	-	2	-	-	-	6	2	3	13
8	1	-	1	8	4	3	45	3	65
9	2	2	-	5	-	53	3	6	71
10	-	-	-	-	-	5	-	1	6
11	-	-	-	-	-	4	1	-	5
12	-	-	1	4	-	3	-	-	8
13+	1	1	-	9	1	2	6	11	31
Totals	17	21	14	87	41	269	224	80	753

DOUBLE TAGGED - DOUBLE RECOVERY

QR	Year Tagged								
	1971	1972	1973	1974	1975	1976	1977	1978	TOTAL
1	8	6	26	50	1	36	10	4	141
2	-	-	3	2	-	-	2	1	8
3	-	-	-	-	1	1	-	-	2
4	9	20	-	14	-	1	23	56	123
5	21	7	7	43	4	6	30	4	122
6	-	-	-	-	-	-	3	-	3
7	-	-	1	-	-	-	1	-	2
8	-	-	1	4	2	-	20	4	31
9	5	1	-	15	-	2	2	5	30
10	-	-	-	-	-	1	-	-	1
11	-	-	-	-	-	1	-	1	2
12	-	-	-	1	-	-	-	-	1
13+	-	-	-	4	-	1	1	2	8
Totals	43	34	38	133	8	49	92	77	474

TABLE 3-10 Continued

DOUBLE TAGGED - SINGLE RECOVERY

QR	Year Tagged								TOTAL
	1971	1972	1973	1974	1975	1976	1977	1978	
1	1	1	7	5	1	4	-	-	19
2	-	-	1	1	-	-	1	-	3
3	-	-	-	-	-	-	1	-	1
4	2	13	-	4	1	2	5	10	37
5	16	2	4	22	5	1	6	3	59
6	-	-	2	-	-	-	1	-	3
7	-	-	-	1	-	-	1	1	3
8	-	1	3	5	-	1	5	3	18
9	4	1	1	11	-	4	1	8	30
10	-	-	-	-	-	1	1	-	2
11	-	-	1	3	-	-	-	2	6
12	-	-	-	1	1	-	-	-	2
13+	-	1	-	5	1	2	3	8	20
Totals	23	19	19	58	9	15	25	35	203

Note: Dash indicates zero value.

values of M (natural mortality plus shedding constant and variable) are compatible with known values of initial population size (number originally marked) and cause only minor changes in age-specific F (Table 3-12 for SS and Table 3-13 for DD). Also, further testing shows that VPA has low sensitivity to the arbitrarily assumed reporting rate of 80% (Table 3-14 for SS and Table 3-15 for DD).

Preliminary examination of the transfer of fish from the east to the west shows 17 fish out of 356 recoveries (Cort and de la Serna, 1993) for the period 1976 through 1991. For west to east movement, the data from MRFISH show 65 of 3,102 total recoveries made in the east. The west to east data are highly variable between years (Table 3-16). When the data were sorted by time elapsed since tagging, the ratio of eastern to western Atlantic recoveries increased with time elapsed (Table 3-17). To obtain more detailed information, the file MRFISH was searched for data on fish tagged from 1960 to 1981 in the west with a single tag. Table 3-5 shows the number tagged by year; Table 3-18 shows the recoveries of single tags in the western Atlantic Ocean by year and quarter after tagging; and Table 3-19 shows the eastern Atlantic Ocean recoveries by year and quarter after tagging. Data for 31 or more quarter years were combined. The longest time to recovery was 18 years (Table 3-5), supporting the belief that the natural mortality rate of bluefin tuna is low.

Data with known times at recovery from the eastern Atlantic Ocean (Table 3-3) were used to estimate fishing mortality rates (F_e). A plot of the natural

TABLE 3-11 Atlantic bluefin tuna tagged in the western Atlantic Ocean from 1971 to 1978. Estimates of annual instantaneous shedding rates using observed catches are shown in A, while B shows the observed catches increased by 20% for nonreporting (i.e., multiplied by 1.2). 5,409 single tagged fish and 4,056 double tagged fish were released. The method used is that of Chapman et al. (1975).

	Time Out Mid Point	Db Db vs Sing	Db Sing vs Sing	Db Db vs Db Sing	Catch Sing	Catch Db Db	Catch Db Sing
A.	0.125	1.5418	0.4574	0.5216	228	141	19
	0.375	1.9300	0.2542	0.4583	22	8	3
	0.625	2.4062	0.0915	0.3570	12	2	1
	0.875	−0.4567	0.2902	0.1601	110	123	37
	1.125	0.0390	0.2340	0.1925	170	122	59
	1.375	0.7989	0.1326	0.2949	12	3	3
	1.625	0.9747	0.1028	0.3444	13	2	3
	1.875	0.2414	0.1089	0.1359	65	31	18
	2.125	0.2699	0.1557	0.1908	71	30	30
	2.375	0.6332	0.1058	0.2919	6	1	2
	2.625	0.2394	0.6134	0.3491	5	2	6
	2.875	0.6232	0.0634	0.2411	8	1	2
B.	Time Out Mid Point	Db Db vs Sing	Db Sing vs Sing	Db Db vs Db Sing	Catch Sing	Catch Db Db	Catch Db Sing
	0.125	1.5417	0.4574	0.5216	274	169	23
	0.375	1.9300	0.2542	0.4583	26	10	4
	0.625	2.4062	0.0915	0.3570	14	2	1
	0.875	−0.4567	0.2902	0.1601	132	148	44
	1.125	0.0390	0.2340	0.1925	204	146	71
	1.375	0.7989	0.1326	0.2949	14	4	4
	1.625	0.9747	0.1028	0.3444	16	2	4
	1.875	0.2414	0.1089	0.1359	78	37	22
	2.125	0.2699	0.1557	0.1908	85	36	36
	2.375	0.6332	0.1058	0.2919	7	1	2
	2.625	0.2394	0.6134	0.3491	6	2	7
	2.875	0.6232	0.0634	0.2411	10	1	2

Sing = Tagged with single tag - recovered with tag
Db Db = Tagged with 2 or more tags - recovered with 2 or more tags
Db Sing = Tagged with 2 or more tags - recovered with one tag
Almost all double tagged fish had only 2 tags.

TABLE 3-12 Atlantic bluefin tuna tagged with a single tag from 1971 to 1978, 5,409 single tagged fish were released. VPA analysis on the reported catches in number of fish by time before recapture stratified by quarter-year intervals. A and B show two different estimates of tag shedding.

A	Time Interval	Initial Pop. Size	Catch	Inst. Fish. Mort.	Inst. Nat. Mort.	Tag Loss Rate	Elapsed Time
	1	5,349.1	228.00	0.1833	0.1400	0.2600	0.25
	2	4,623.4	22.00	0.0200	0.1400	0.2600	0.25
	3	4,162.5	12.00	0.0121	0.1400	0.2600	0.25
	4	3,755.1	110.00	0.1250	0.1400	0.2600	0.25
	5	3,293.2	170.00	0.2229	0.1400	0.2600	0.25
	6	2,818.3	12.00	0.0179	0.1400	0.2600	0.25
	7	2,538.8	13.00	0.0215	0.1400	0.2600	0.25
	8	2,284.8	65.00	0.1213	0.1400	0.2600	0.25
	9	2,005.6	71.00	0.1515	0.1400	0.2600	0.25
	10	1,747.3	6.00	0.0145	0.1400	0.2600	0.25
	11	1,575.3	5.00	0.0133	0.1400	0.2600	0.25
	12	1,420.7	8.00	0.0237	0.1400	0.2600	0.25
	13	930.0	31.00	0.0138	0.1400	0.2600	12.25
	TOTALS		753.0	0.2317	2.9150	3.1850	15.25
B	Time Interval	Initial Pop. Size	Catch	Inst. Fish. Mort.	Inst. Nat. Mort.	Tag Loss Rate	Elapsed Time
	1	5,357.8	228.00	0.1884	0.1400	0.5000	0.25
	2	4,355.5	22.00	0.0215	0.1400	0.4000	0.25
	3	3,785.1	12.00	0.0133	0.1400	0.3000	0.25
	4	,3379.6	110.00	0.1381	0.1400	0.2000	0.25
	5	2,998.8	170.00	0.2437	0.1400	0.2000	0.25
	6	2,591.7	12.00	0.0194	0.1400	0.2000	0.25
	7	2,369.0	13.00	0.0230	0.1400	0.2000	0.25
	8	2,163.4	65.00	0.1274	0.1400	0.2000	0.25
	9	1,924.8	71.00	0.1570	0.1400	0.2000	0.25
	10	1,699.9	6.00	0.0148	0.1400	0.2000	0.25
	11	1,555.6	5.00	0.0136	0.1400	0.2000	0.25
	12	1,424.0	8.00	0.0237	0.1400	0.2000	0.25
	13	990.0	31.00	0.0110	0.1400	0.2000	12.25
	TOTALS		753.0	0.2460	2.8850	2.4500	15.25

TABLE 3-13 Atlantic bluefin tuna tagged with two or more tags from 1971 to 1978, 4,056 fish were released. VPA analysis on the reported catches, with two tags remaining, in number of fish by time before recapture stratified by quarter-year intervals. A and B show two different estimates of tag shedding.

A Time Interval	Initial Pop Size	Catch	Inst. Fish. Mort.	Inst. Nat. Mort.	Tag Loss Rate	Elapsed Time
1	3,955.0	141.00	0.1576	0.1400	0.5200	0.25
2	3,223.9	8.00	0.0108	0.1400	0.5200	0.25
3	2,726.1	2.00	0.0032	0.1400	0.5200	0.25
4	2,309.6	123.00	0.2379	0.1400	0.5200	0.25
5	1,845.2	122.00	0.2974	0.1400	0.5200	0.25
6	1,452.5	3.00	0.0090	0.1400	0.5200	0.25
7	1,228.8	2.00	0.0072	0.1400	0.5200	0.25
8	1,040.0	31.00	0.1314	0.1400	0.5200	0.25
9	853.3	30.00	0.1555	0.1400	0.5200	0.25
10	695.9	1.00	0.0063	0.1400	0.5200	0.25
11	589.2	2.00	0.0148	0.1400	0.5200	0.25
12	497.7	1.00	0.0087	0.1400	0.5200	0.25
13	255.0	8.00	0.0214	0.1400	0.5200	12.25
TOTALS		474.00	0.2599	3.6950	6.3700	15.25

B Time Interval	Initial Pop Size	Catch	Inst. Fish. Mort.	Inst. Nat. Mort.	Tag Loss Rate	Elapsed Time
1	3,958.7	141.00	0.1671	0.1400	1.0000	0.25
2	2,855.2	8.00	0.0127	0.1400	0.8000	0.25
3	2,250.1	2.00	0.0038	0.1400	0.6000	0.25
4	1,868.3	123.00	0.2916	0.1400	0.4000	0.25
5	1,517.6	122.00	0.3590	0.1400	0.4000	0.25
6	1,212.1	3.00	0.0105	0.1400	0.4000	0.25
7	1,056.3	2.00	0.0081	0.1400	0.4000	0.25
8	921.0	31.00	0.1466	0.1400	0.4000	0.25
9	775.7	30.00	0.1689	0.1400	0.4000	0.25
10	649.8	1.00	0.0066	0.1400	0.4000	0.25
11	566.8	2.00	0.0151	0.1400	0.4000	0.25
12	493.3	1.00	0.0087	0.1400	0.4000	0.25
13	285.0	8.00	0.0156	0.1400	0.4000	12.25
TOTALS		474.00	0.2997	3.6350	4.9000	15.25

TABLE 3-14 Atlantic bluefin tuna tagged with a single tag from 1971 to 1978, 5,409 fish were released. VPA analysis on the reported catches in number of fish, increased by 20% (i.e. multiplied by 1.2) to account for assumed non-reporting of tags, by time before recapture stratified by quarter-year intervals. A and B show two different estimates of tag shedding.

A	Time Interval	Initial Pop. Size	Catch	Inst. Fish. Mort.	Inst. Nat. Mort.	Tag Loss Rate	Elapsed Time
	1	5,347.9	273.60	0.2208	0.1400	0.2600	0.25
	2	4,579.1	26.40	0.0243	0.1400	0.2600	0.25
	3	4,118.3	14.40	0.0148	0.1400	0.2600	0.25
	4	3,712.6	132.00	0.1521	0.1400	0.2600	0.25
	5	3,233.9	204.00	0.2742	0.1400	0.2600	0.25
	6	2,732.3	14.40	0.0221	0.1400	0.2600	0.25
	7	2,458.7	15.60	0.0267	0.1400	0.2600	0.25
	8	2,209.9	78.00	0.1512	0.1400	0.2600	0.25
	9	1,925.4	85.20	0.1903	0.1400	0.2600	0.25
	10	1,661.3	7.20	0.0182	0.1400	0.2600	0.25
	11	1,496.4	6.00	0.0169	0.1400	0.2600	0.25
	12	1,348.2	9.60	0.0301	0.1400	0.2600	0.25
	13	877.0	37.20	0.0177	0.1400	0.2600	12.25
	TOTALS		903.60	0.2854	2.9150	3.1850	15.25

B	Time Interval	Initial Pop. Size	Catch	Inst. Fish. Mort.	Inst. Nat. Mort.	Tag Loss Rate	Elapsed Time
	1	5,355.8	273.60	0.2272	0.1400	0.5000	0.25
	2	4,311.9	26.40	0.0264	0.1400	0.4000	0.25
	3	3,742.6	14.40	0.0163	0.1400	0.3000	0.25
	4	3,339.1	132.00	0.1683	0.1400	0.2000	0.25
	5	2,940.6	204.00	0.3001	0.1400	0.2000	0.25
	6	2,505.7	14.40	0.0240	0.1400	0.2000	0.25
	7	2,287.8	15.60	0.0285	0.1400	0.2000	0.25
	8	2,086.4	78.00	0.1591	0.1400	0.2000	0.25
	9	1,841.7	85.20	0.1979	0.1400	0.2000	0.25
	10	1,609.9	7.20	0.0188	0.1400	0.2000	0.25
	11	1,471.8	6.00	0.0169	0.1400	0.2000	0.25
	12	1,346.2	9.60	0.0298	0.1400	0.2000	0.25
	13	930.0	37.20	0.0142	0.1400	0.2000	12.25
	TOTALS		903.60	0.3033	2.8850	2.4500	15.25

TABLE 3-15 Atlantic bluefin tuna tagged with a two or more tags from 1971 to 1978, 4,056 fish were released. VPA analysis on the reported catches in number of fish with two tags remaining, increased by 20% (i.e. multiplied by 1.2) to account for assumed non-reporting of tags, by time before recapture stratified by quarter-year intervals. A and B show two different estimates of tag shedding.

A Time Interval	Initial Pop. Size	Catch	Inst. Fish. Mort.	Inst. Nat. Mort.	Tag Loss Rate	Elapsed Time
1	3,961.0	169.20	0.1897	0.1400	0.5200	0.25
2	3,203.0	9.60	0.0130	0.1400	0.5200	0.25
3	2,707.0	2.40	0.0038	0.1400	0.5200	0.25
4	2,293.0	147.60	0.2892	0.1400	0.5200	0.25
5	1,808.7	146.40	0.3670	0.1400	0.5200	0.25
6	1,399.1	3.60	0.0111	0.1400	0.5200	0.25
7	1,183.0	2.40	0.0087	0.1400	0.5200	0.25
8	1,000.9	37.20	0.1646	0.1400	0.5200	0.25
9	814.4	36.00	0.1964	0.1400	0.5200	0.25
10	657.5	1.20	0.0078	0.1400	0.5200	0.25
11	556.4	2.40	0.0188	0.1400	0.5200	0.25
12	469.5	1.20	0.0111	0.1400	0.5200	0.25
13	240.0	9.60	0.0275	0.1400	0.5200	12.25
TOTALS		568.80	0.3203	3.6950	6.3700	15.25

B Time Interval	Initial Pop. Size	Catch	Inst. Fish. Mort.	Inst. Nat. Mort.	Tag Loss Rate	Elapsed Time
1	3,962.3	169.20	0.2010	0.1400	1.0000	0.25
2	2,833.7	9.60	0.0151	0.1400	0.8000	0.25
3	2,231.8	2.40	0.0047	0.1400	0.6000	0.25
4	1,852.7	147.60	0.3557	0.1400	0.4000	0.25
5	1,481.0	146.40	0.4463	0.1400	0.4000	0.25
6	1,157.3	3.60	0.0133	0.1400	0.4000	0.25
7	1,007.8	2.40	0.0102	0.1400	0.4000	0.25
8	878.3	37.20	0.1851	0.1400	0.4000	0.25
9	732.7	36.00	0.2156	0.1400	0.4000	0.25
10	606.6	1.20	0.0084	0.1400	0.4000	0.25
11	528.9	2.40	0.0194	0.1400	0.4000	0.25
12	459.8	1.20	0.0111	0.1400	0.4000	0.25
13	265.0	9.60	0.0203	0.1400	0.4000	12.25
TOTALS		568.80	0.3715	3.6350	4.9000	15.25

TABLE 3-16 Atlantic bluefin tuna tagged in the western Atlantic Ocean with single, double, and multiple tags. Number of fish recaptured, which retained tags, by ocean area, western and eastern, and the percent of known recoveries made in the east.

Year Tagged	Unknown Area	West Area	East Area	Total Known	Percent Known East
1965	10	238	19	257	7.393
1966	47	1,107	18	1,125	1.600
1967	4	181	3	184	1.630
1968	3	113	0	113	0.000
1969	10	102	2	104	1.923
1970	19	162	9	171	5.263
1971	5	78	0	78	0.000
1972	1	72	1	73	1.370
1973	3	67	1	68	1.471
1974	16	262	0	262	0.000
1975	3	55	0	55	0.000
1976	85	247	1	248	0.403
1977	6	331	4	335	1.194
1978	2	185	5	190	2.632
1979	3	49	1	50	2.000
1980	0	266	2	268	0.746
1981	0	67	1	68	1.471
TOTALS	217	3,582	67	3,649	1.836

TABLE 3-17 Atlantic bluefin tuna tagged in the western Atlantic Ocean with a single tag. Ratio of eastern Atlantic recoveries to western recoveries as a function of time before recapture.

West	East	Ratio	Time Out (days)	Interval Length (years)
1,356	7	0.0052	0 TO 180	0.5
1,286	36	0.0280	181 TO 545	1.0
277	15	0.0542	546 TO 910	1.0
54	1	0.0185	911 TO 1,275	1.0
64	6	0.0938	1,276 OR MORE	15.0

TABLE 3-18 Atlantic bluefin tuna tagged with a single tag in the western Atlantic Ocean and recovered from the western Atlantic Ocean by year and by number of quarters out.

Year	Total	Recoveries by Quarter														
		0	1	2	3	4	5	6	7	8	9	10	11	12	13	14
1960	3	-	-	-	-	-	-	-	-	1	-	1	-	-	-	-
1961	7	-	-	-	1	1	-	-	2	1	-	-	2	-	-	-
1962	4	-	-	-	2	2	-	-	-	-	-	-	-	-	-	-
1963	10	7	-	-	-	-	-	-	1	-	-	-	-	-	-	-
1964	117	92	-	-	15	9	1	-	-	-	-	-	-	-	-	-
1965	238	159	-	-	23	23	1	-	16	13	-	-	1	1	-	-
1966	1,104	477	7	-	261	293	-	-	32	16	-	-	3	5	-	-
1967	181	95	1	-	33	25	1	-	12	3	-	-	9	1	-	-
1968	76	58	2	2	5	5	-	-	3	1	-	-	-	-	-	-
1969	97	14	-	-	54	15	-	-	8	1	-	-	1	-	-	1
1970	160	48	2	-	68	25	-	-	2	6	-	-	7	-	-	-
1971	17	2	-	-	5	6	-	-	2	1	-	-	-	-	-	-
1972	20	3	-	-	4	7	2	1	1	1	-	-	-	1	-	-
1973	13	4	-	-	3	4	-	-	1	-	-	-	1	-	-	-
1974	77	16	-	-	31	7	1	-	10	-	-	-	6	-	1	3
1975	39	20	-	-	9	5	-	-	4	-	-	-	-	-	-	-
1976	201	80	3	-	43	2	5	-	48	10	4	1	3	-	1	-
1977	219	37	4	6	94	19	5	1	45	1	1	-	-	1	-	-
1978	78	18	1	2	33	1	1	2	5	4	-	-	-	4	-	-
1979	46	5	5	3	12	6	1	1	7	-	-	-	1	-	-	-
1980	266	141	-	3	62	35	-	-	-	6	-	3	-	-	-	3
1981	64	53	2	-	2	-	-	1	1	-	1	-	-	-	-	-
Total	3,037	1,329	27	16	762	490	18	6	200	65	6	5	34	13	2	4

Note: Dash indicates zero value.

15	16	17	18	19	20	21	22	23	24	25	26	27	28	29	30	31+
-	-	-	-	-	-	-	-	-	-	-	-	-	-	-	-	1
-	-	-	-	-	-	-	-	-	-	-	-	-	-	-	-	-
-	-	-	-	-	-	-	-	-	-	-	-	-	-	-	-	-
-	-	-	-	-	-	-	-	-	-	-	-	-	-	-	-	-
-	-	-	-	-	-	-	-	-	-	-	-	-	-	-	-	-
-	-	-	-	-	-	-	-	-	-	-	-	-	-	-	-	1
1	4	-	-	-	-	-	-	1	1	-	-	-	-	-	-	3
-	-	-	-	-	-	-	-	-	-	-	-	-	-	-	-	1
-	-	-	-	-	-	-	-	-	-	-	-	-	-	-	-	-
-	1	-	-	-	-	-	-	1	-	-	-	-	-	-	-	1
1	1	-	-	-	-	-	-	-	-	-	-	-	-	-	-	-
-	-	-	-	1	-	-	-	-	-	-	-	-	-	-	-	-
-	-	-	-	-	-	-	-	-	-	-	-	-	-	-	-	-
-	-	-	-	-	-	-	-	-	-	-	-	-	-	-	-	-
3	-	-	-	-	-	-	-	-	-	-	-	-	-	-	-	2
-	-	-	-	-	-	-	-	-	1	-	-	-	-	-	-	-
-	-	-	-	1	-	-	-	-	-	-	-	-	-	-	-	-
-	-	-	-	-	-	-	-	-	-	-	-	-	1	-	-	4
2	2	1	-	-	-	-	-	1	-	-	-	-	-	-	-	1
3	-	-	-	-	1	-	-	1	-	-	-	-	-	-	-	-
-	1	-	1	-	1	-	-	-	1	-	-	-	3	1	-	5
-	1	-	-	-	-	-	-	-	1	-	-	-	-	-	-	2
10	10	1	1	2	2	-	-	4	4	-	-	-	4	1	-	21

TABLE 3-19 Atlantic bluefin tuna tagged with a single tag in the western Atlantic Ocean and recovered from the eastern Atlantic Ocean by year and by number of quarters out.

		Recoveries by Quarter														
Year	Total	0	1	2	3	4	5	6	7	8	9	10	11	12	13	14
1960	2	-	-	-	-	-	-	-	-	2	-	-	-	-	-	-
1961	2	-	2	-	-	-	-	-	-	-	-	-	-	-	-	-
1962	1	1	-	-	-	-	-	-	-	-	-	-	-	-	-	-
1965	19	-	-	-	5	7	-	1	4	1	-	-	1	-	-	-
1966	18	1	-	-	10	2	-	-	1	2	-	-	-	-	-	-
1967	3	1	-	-	-	-	1	-	1	-	-	-	-	-	-	-
1969	2	-	-	-	-	-	-	1	-	-	-	-	-	-	-	-
1970	9	-	2	-	5	2	-	-	-	-	-	-	-	-	-	-
1972	1	-	-	-	-	-	1	-	-	-	-	-	-	-	-	-
1973	1	-	-	-	-	-	1	-	-	-	-	-	-	-	-	-
1977	2	-	-	-	-	1	-	-	-	1	-	-	-	-	-	-
1978	1	-	-	-	-	-	-	-	1	-	-	-	-	-	-	-
1979	1	-	-	-	1	-	-	-	-	-	-	-	-	-	-	-
1980	2	-	-	-	-	-	-	-	-	-	-	-	-	-	-	-
1981	1	-	-	-	-	-	-	-	-	-	-	-	-	-	-	-
Total	65	3	4	-	21	12	3	1	8	6	-	-	1	-	-	-

Note: Dash indicates zero value.

logarithm of the recovery frequency by year against the time until recovery shows a steep decline rate (F_e+M+T_e) of 1.03 (Table 3-20). If $M = 0.40$ (0.14 [natural] + 0.26 [shedding]) and T_e is between 0.01 and 0.05, the F_e would be about 0.60. A similar plot of the data in Table 3-5, summed over 1960 to 1981 and leaving out year 0, gives a decline rate of 1.31 (Table 3-21) for the western Atlantic Ocean. That would give a value of F_w of about 0.90. Application of the Chapman-Robson method (Chapman and Robson, 1960) gives slightly lower values for the east (with F_e varying between 0.36 and 0.63, depending on which year of recover set is used) and somewhat higher values for the west (F_w varying between 0.58 and 1.14). Estimates of F from the methods discussed in this paragraph are likely to be erroneously high owing, to a large extent, to effects on the results from the failure of the assumption of a constant F over time, which is implicit in such methods.

Data on the frequency by year against time (280 total recoveries) in Table 3-3 (eastern Atlantic Ocean) were expanded by the ratio 356 to 280 to obtain an age-frequency data set for VPA analysis. Backward calculation was used for six different values of $F_e(5)$: 0.5, 0.3, 0.1, 0.05, 0.016, and 0.01. Results for the eastern data are shown in Table 3-22. Similarly, data on frequency by year against time (western Atlantic Ocean) in Table 3-5 were used to do backward VPA analysis for five different values of $F_w(6)$. The five values of $F_w(6)$ were

15	16	17	18	19	20	21	22	23	24	25	26	27	28	29	30	31+
-	-	-	-	-	-	-	-	-	-	-	-	-	-	-	-	-
-	-	-	-	-	-	-	-	-	-	-	-	-	-	-	-	-
-	-	-	-	-	-	-	-	-	-	-	-	-	-	-	-	-
-	1	-	-	-	-	-	-	-	-	-	-	-	-	-	-	1
-	-	-	-	-	-	-	-	-	-	-	-	-	-	-	-	-
-	-	-	-	-	-	-	-	-	-	-	-	-	1	-	-	-
-	-	-	-	-	-	-	-	-	-	-	-	-	-	-	-	-
-	-	-	-	-	-	-	-	-	-	-	-	-	-	-	-	-
-	-	-	-	-	-	-	-	-	-	-	-	-	-	-	-	-
-	-	-	-	-	-	-	-	-	-	-	-	-	-	-	-	-
-	-	-	-	-	-	-	-	-	-	-	-	-	-	-	-	-
-	-	-	-	-	-	-	-	1	-	-	-	-	-	-	-	1
-	-	-	-	1	-	-	-	-	-	-	-	-	-	-	-	-
-	1	-	-	1	-	-	-	1	-	-	-	-	1	-	-	2

estimated by assuming five different values (1,600, 1,000, 500, 250, and 100) for the population size at the end of year 6. The value of $F_w(7)$ was estimated from the equation $C = NF/Z$, where C is the catch for age 7+ year-old fish, and N is the guess of the population size. Results for the western data are given in Table 3-23. In Tables 3-22 and 3-23 the natural mortality rate includes the shedding rate of 0.26 and the actual mortality of 0.14. Some slight errors in the F_w estimates occur because eastern Atlantic recoveries, although few, are part of the total recoveries used in the analysis.

Estimates of P_e and P_w were obtained from the ratio of $N_e(1)/5{,}663$ in Table 3-22 and $N_w(1)/20{,}951$ in Table 3-23. Estimates of F_e, Z_e, F_w and Z_w were obtained by dividing totals of the appropriate rates by the total number of time units to obtain average values. The estimates are as follows:

Eastern Atlantic Ocean

(1) $P_e = 0.0978$ $F_e = 0.5995$ $Z_e = 0.9995$
(2) $P_e = 0.1037$ $F_e = 0.4739$ $Z_e = 0.8739$
(3) $P_e = 0.1330$ $F_e = 0.2606$ $Z_e = 0.6606$
(4) $P_e = 0.1766$ $F_e = 0.1636$ $Z_e = 0.5636$
(5) $P_e = 0.3616$ $F_e = 0.0656$ $Z_e = 0.4656$
(6) $P_e = 0.5244$ $F_e = 0.0430$ $Z_e = 0.4043$

TABLE 3-20 Atlantic bluefin tuna tagged in the eastern Atlantic Ocean and recovered in the eastern Atlantic Ocean versus time (years) before recapture and a regression analysis of the natural logarithm of catch number against years out (Cort and LaSerna, 1993).

Interval	Catch Number	Elapsed Time	Nat. Log of Catch
1	152	0.5	5.02388
2	84	1.0	4.43082
3	29	2.0	3.36730
4	11	3.0	2.39790
5	4	4.0	1.38629

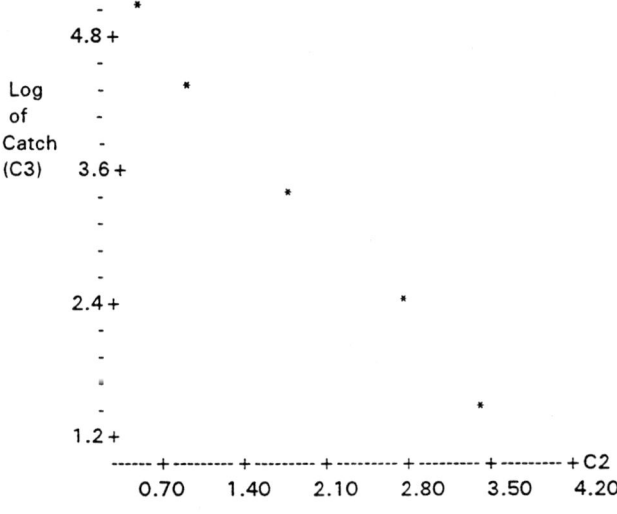

NOTE: The regression equation is C3 = 5.49 - 1.03 C2

Predictor	Coef	Stdev	t-ratio	p
Constant	5.48700	0.04226	129.84	0.000
C2	-1.03131	0.01718	-60.03	0.000
s = 0.04920	R-sq = 99.9%	R-sq(adj) = 99.9%		

TABLE 3-21 Atlantic bluefin tuna tagged in the western Atlantic Ocean from 1960 to 1981 with a single tag and recovered in the western Atlantic Ocean vs. years before recapture and a regression analysis of the natural logarithm of catch number against years out (data in file MRFISH provided by NMFS).

Year	Catch	Log Catch
1	1,386	7.23418
2	309	5.73334
3	59	4.07754
4	26	3.25810
5	7	1.94591

```
                  -
                  -  *
                  -
                  -
             6.0 +
                  -        *
                  -
         Log      -
          of      -
         Catch 4.0 +
                  -                  *
                  -
                  -                         *
                  -
                  -
             2.0 +                                       *
                  -
                  -
                  + --------- + --------- + --------- + --------- + --------- + ------ C1
                0.80       1.60       2.40       3.20       4.00       4.80

                                      Years
```

NOTE: The regression equation is C3 = 8.37 - 1.31 C1

Predictor	Coef	Stdev	t-ratio	p
Constant	8.3653	0.2664	31.40	0.000
C1	-1.30518	0.08032	-16.25	0.001

$s = 0.2540$ R-sq = 98.9%   R-sq(adj) = 98.5%

Analysis of Variance

SOURCE	DF	SS	MS	F	p
Regression	1	17.035	17.035	264.05	0.001
Error	3	0.194	0.065		
Total	4	17.228			

TABLE 3-22 Atlantic bluefin tuna. VPA analysis on fish tagged in the eastern Atlantic Ocean (near Spain) from 1976 to 1991. Six different solutions.

Time	Initial pop	Catch	Fish. Mort.	Nat. Mort.	Shed Rate	Time Units
1	553.8	193.00	0.9608	0.1400	0.2600	0.50
2	280.5	107.00	0.6047	0.1400	0.2600	1.00
3	102.7	37.00	0.5605	0.1400	0.2600	1.00
4	39.3	14.00	0.5522	0.1400	0.2600	1.00
5	15.2	5.00	0.5000	0.1400	0.2600	1.00
TOTALS		356.00	2.6978	0.6300	1.1700	4.50

Time	Initial pop	Catch	Fish. Mort.	Nat. Mort.	Shed Rate	Time Units
1	587.3	193.00	0.8922	0.1400	0.2600	0.50
2	307.8	107.00	0.5351	0.1400	0.2600	1.00
3	120.8	37.00	0.4558	0.1400	0.2600	1.00
4	51.3	14.00	0.3954	0.1400	0.2600	1.00
5	23.2	5.00	0.3001	0.1400	0.2600	1.00
TOTALS		356.00	2.1325	0.6300	1.1700	4.50

Time	Initial Pop	Catch	Fish. Mort.	Nat. Mort.	Shed Rate	Time Units
1	753.3	193.00	0.6599	0.1400	0.2600	0.50
2	443.4	107.00	0.3419	0.1400	0.2600	1.00
3	211.2	37.00	0.2370	0.1400	0.2600	1.00
4	111.7	14.00	0.1640	0.1400	0.2600	1.00
5	63.5	5.00	0.0999	0.1400	0.2600	1.00
TOTALS		356.00	1.1729	0.6300	1.1700	4.50

TABLE 3-22 Continued

Time	Initial Pop	Catch	Fish. Mort.	Nat. Mort.	Shed Rate	Time Units
1	1,000.3	193.00	0.4765	0.1400	0.2600	0.50
2	645.4	107.00	0.2226	0.1400	0.2600	1.00
3	346.3	37.00	0.1381	0.1400	0.2600	1.00
4	202.2	14.00	0.0874	0.1400	0.2600	1.00
5	124.2	5.00	0.0499	0.1400	0.2600	1.00
TOTALS		356.00	0.7363	0.6300	1.1700	4.50

Time	Initial Pop	Catch	Fish. Mort.	Nat. Mort.	Shed Rate	Time Units
1	2,047.9	193.00	0.2193	0.1400	0.2600	0.50
2	1,502.6	107.00	0.0902	0.1400	0.2600	1.00
3	920.4	37.00	0.0499	0.1400	0.2600	1.00
4	586.9	14.00	0.0294	0.1400	0.2600	1.00
5	382.0	5.00	0.0160	0.1400	0.2600	1.00
TOTALS		356.00	0.2952	0.6300	1.1700	4.50

Time	Initial Pop	Catch	Fish. Mort.	Nat. Mort.	Shed Rate	Time Units
1	2,969.6	193.00	0.1485	0.1400	0.2600	0.50
2	2,257.3	107.00	0.0591	0.1400	0.2600	1.00
3	1,426.4	37.00	0.0319	0.1400	0.2600	1.00
4	926.1	14.00	0.0185	0.1400	0.2600	1.00
5	609.4	5.00	0.0099	0.1400	0.2600	1.00
TOTALS		356.00	0.1936	0.6300	1.1700	4.50

TABLE 3-23 Atlantic bluefin tuna. VPA analysis on fish tagged in the western Atlantic Ocean (near USA) from 1960 to 1981. Five different solutions.

Time	Initial Pop	Catch	Fish. Mort.	Nat. Mort.	Shed Rate	Time Units
1	19,162.9	1,471.00	0.1768	0.1400	0.2600	0.50
2	14,361.5	1,386.00	0.1241	0.1400	0.2600	1.00
3	8,503.7	309.00	0.0450	0.1400	0.2600	1.00
4	5,449.3	59.00	0.0133	0.1400	0.2600	1.00
5	3,604.6	26.00	0.0087	0.1400	0.2600	1.00
6	2,395.3	7.00	0.0035	0.1400	0.2600	1.00
7	1,600.0	44.00	0.0113	0.1400	0.2600	10.00
TOTALS		3,302.00	0.3961	2.1700	4.0300	15.50

Time	Initial Pop	Catch	Fish. Mort.	Nat. Mort.	Shed Rate	Time Units
1	13,739.9	1,471.00	0.2507	0.1400	0.2600	0.50
2	9,924.0	1,386.00	0.1845	0.1400	0.2600	1.00
3	5,531.6	309.00	0.0700	0.1400	0.2600	1.00
4	3,457.1	59.00	0.0209	0.1400	0.2600	1.00
5	2,269.4	26.00	0.0139	0.1400	0.2600	1.00
6	1,500.3	7.00	0.0056	0.1400	0.2600	1.00
7	1,000.0	44.00	0.0184	0.1400	0.2600	10.00
TOTALS		3,302.00	0.6044	2.1700	4.0300	15.50

Time	Initial Pop	Catch	Fish. Mort.	Nat. Mort.	Shed Rate	Time Units
1	9,218.3	1,471.00	0.3859	0.1400	0.2600	0.50
2	6,223.0	1,386.00	0.3114	0.1400	0.2600	1.00
3	3,055.1	309.00	0.1305	0.1400	0.2600	1.00
4	1,797.4	59.00	0.0404	0.1400	0.2600	1.00
5	1,157.1	26.00	0.0276	0.1400	0.2600	1.00
6	754.5	7.00	0.0114	0.1400	0.2600	1.00
7	500.0	44.00	0.0386	0.1400	0.2600	10.00
TOTALS		3,302.00	1.1003	2.1700	4.0300	15.50

TABLE 3-23 Continued

Time	Initial Pop	Catch	Fish. Mort.	Nat. Mort.	Shed Rate	Time Units
1	6,945.2	1,471.00	0.5296	0.1400	0.2600	0.50
2	4,363.3	1,386.00	0.4771	0.1400	0.2600	1.00
3	1,815.0	309.00	0.2293	0.1400	0.2600	1.00
4	967.3	59.00	0.0768	0.1400	0.2600	1.00
5	600.5	26.00	0.0539	0.1400	0.2600	1.00
6	381.4	7.00	0.0224	0.1400	0.2600	1.00
7	250.0	44.00	0.0854	0.1400	0.2600	10.00
TOTALS		3,302.00	1.9787	2.1700	4.0300	15.50

Time	Initial Pop	Catch	Fish. Mort.	Nat. Mort.	Shed Rate	Time Units
1	5,570.1	1,471.00	0.6844	0.1400	0.2600	0.50
2	3,238.9	1,386.00	0.7082	0.1400	0.2600	1.00
3	1,069.4	309.00	0.4243	0.1400	0.2600	1.00
4	468.9	59.00	0.1646	0.1400	0.2600	1.00
5	266.6	26.00	0.1253	0.1400	0.2600	1.00
6	157.7	7.00	0.0554	0.1400	0.2600	1.00
7	100.0	44.00	0.3143	0.1400	0.2600	10.00
TOTALS		3,302.00	4.9628	2.1700	4.0300	15.50

TABLE 3-24 Estimates of transfer rates from east to west in the Atlantic Ocean, using tagging data and assuming a natural mortality rate of 0.14, a tag shedding rate of 0.26, and an eastern catch of 339 and a western catch of 17, with a ratio of 0.0501.

	$P_e = 0.5244$ $F_e = 0.0430$	$P_e = 0.3616$ $F_e = 0.0656$	$P_e = 0.1330$ $F_e = 0.2606$
$P_w = 0.9147$ $F_w = 0.0256$	$T_e = 0.0206$	$T_e = 0.0216$	$T_e = 0.0316$
$P_w = 0.6558$ $F_w = 0.0390$	$T_e = 0.0194$	$T_e = 0.0204$	$T_e = 0.0298$
$P_w = 0.4400$ $F_w = 0.0710$	$T_e = 0.0171$	$T_e = 0.0179$	$T_e = 0.0262$
$P_w = 0.3315$ $F_w = 0.1277$	$T_e = 0.0141$	$T_e = 0.0148$	$T_e = 0.0217$
$P_w = 0.2659$ $F_w = 0.3202$	$T_e = 0.0096$	$T_e = 0.0101$	$T_e = 0.0147$

P_w = Fraction of captured tags which are reported in the west.

P_e = Fraction of captured tags which are reported in the east

F_w = Annual instantaneous fishing mortality rate in the west.

F_e = Annual instantaneous fishing mortality rate in the east.

T_e = Annual instantaneous rate of transfer from the east to the west.

TABLE 3-25 Estimates of transfer rates from west to east in the Atlantic Ocean, using tagging data and assuming a natural mortality rate of 0.14, a tag shedding rate of 0.26, and a western catch of 3037 and an eastern catch of 65, with a ratio of 0.0214.

	$P_w = 0.9147$ $F_w = 0.0256$	$P_w = 0.4400$ $F_w = 0.0710$	$P_w = 0.2659$ $F_w = 0.3202$
$P_e = 0.5244$ $F_e = 0.0430$	$T_w = 0.0099$	$T_w = 0.0131$	$T_w = 0.0358$
$P_e = 0.3616$ $F_e = 0.0656$	$T_w = 0.0098$	$T_w = 0.0130$	$T_w = 0.0356$
$P_e = 0.1330$ $F_e = 0.2606$	$T_w = 0.0096$	$T_w = 0.0127$	$T_w = 0.0347$
$P_e = 0.1037$ $F_e = 0.4739$	$T_w = 0.0089$	$T_w = 0.0119$	$T_w = 0.0324$
$P_e = 0.0978$ $F_e = 0.5995$	$T_w = 0.0085$	$T_w = 0.0114$	$T_w = 0.0311$

P_w = Fraction of captured tags which are reported in the west.

P_e = Fraction of captured tags which are reported in the east.

F_w = Annual instantaneous fishing mortality rate in the west.

F_e = Annual instantaneous fishing mortality rate in the east.

T_w = Annual instantaneous rate of transfer from the west to the east.

Western Atlantic Ocean

(1) $P_w = 0.9147$ $F_w = 0.0256$ $Z_w = 0.4256$
(2) $P_w = 0.6558$ $F_w = 0.0390$ $Z_w = 0.4390$
(3) $P_w = 0.4400$ $F_w = 0.0710$ $Z_w = 0.4710$
(4) $P_w = 0.3315$ $F_w = 0.1277$ $Z_w = 0.5277$
(5) $P_w = 0.2659$ $F_w = 0.3202$ $Z_w = 0.7202$

These estimates were used to compute a variety of transfer rates from east to west (Table 3-24) and from west to east (Table 3-25). *Not all combinations of P values are used in Tables 3-24 and 3-25 because other results were inconsistent with VPA estimates of Chapter 4.*

DISCUSSION

Shedding rates appear to be 0.20 or higher (Table 3-11). Rates estimated by comparing double-double returns to double-single returns should be the most robust estimates, because the two groups originated from the same group of tags and should have been exposed to similar histories. Rates appear to be higher in the first several quarters. Two different sets of shedding rates were used in the VPA analysis (Tables 3-12 to 3-15).

The results of this reanalysis of tagging data provide quantitative confirmation of the empirical results on transfer rates presented earlier in Chapter 3. Although the committee was unable to obtain variance estimates for these transfer rates due to time constraints, results are robust with respect to reporting rate, the parameter most likely to affect the results. Thus, the results are expected to be statistically significant. However, movement of fish from west to east seems to have varied from year to year (Table 3-16). Movement from west to east also seems to have a cumulative effect, with the percent caught in the east increasing as a function of the amount of time the tags were attached to the fish (Table 3-17).

Annual transfer rates estimated with the model (equations 1 and 2) are about 0.01 for west to east, about 0.02 for east to west, and 0.03 overall. These rates depend to a great extent on the nonreporting rate for captured tagged fish. The VPA analysis on the Spanish data is especially indicative of a high nonreporting rate. The nonreporting rate for the west could also be high. Little can be said about the estimate of natural mortality from these data. However, even though there is good reason to be skeptical about the nonreporting rate and natural mortality rate, the transfer rates are almost certainly low, because the number of tags recovered in the other area is small compared to the total tags recovered.

Information is not available to determine whether a transfer from one side

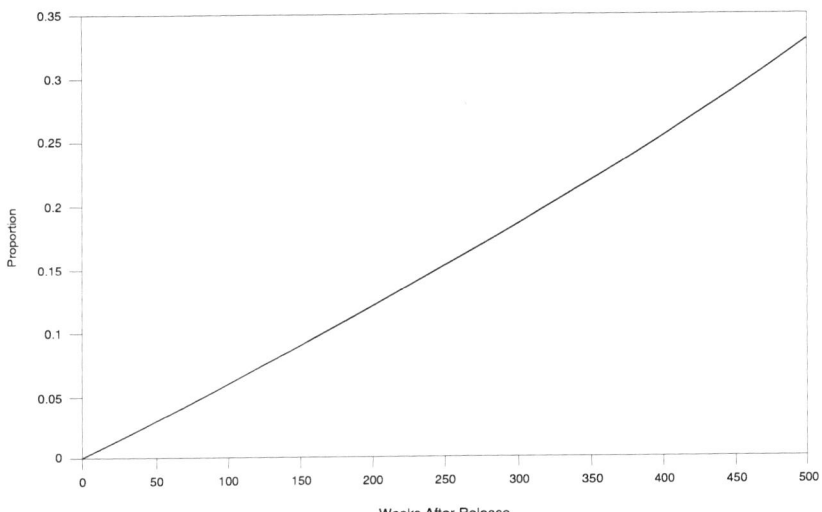

FIGURE 3-1 Proportion of tagged bluefin tuna in area 2 (eastern Atlantic Ocean) that were originally released in area 1 (western Atlantic Ocean) for a hypothetical population with $T_w=0.03$, $F=0.1$, and $M=0.14$.

to the other is permanent. The consequences of returning would give a different meaning to the data. If fish that transfer each year do not return, even a low transfer rate will cause a significant portion of the spawning-age fish produced as recruits on one side to be on the other side when they are sexually mature. Figure 3-1 shows the increasing proportion of tagged fish in area 2 that were originally released in area 1 for a hypothetical population with $T=0.03$, $F=0.1$, and $M=0.14$. After 300 weeks (about 6 years), the proportion is up to 0.20 even though the annual transfer rate is only 0.03; that is, 20% of the fish on one side of the Atlantic Ocean will have originated from the other side. This example shows that small annual transfer rates can have a large effect on population dynamics over a relatively short time period.

Because none of the fish tagged in the eastern Atlantic Ocean were from the Mediterranean Sea, caution should be exercised in interpreting the east to west movement rate.

Estimates of P_e and P_w obtained from Tables 3-22 and 3-23 were calculated with the assumption of no type I (initial) tagging mortality. If some, or most, of the differences between the VPA estimates of age 1 fish in the east and west result from type I mortality, the estimates of P_e and P_w would be incorrect. However, it is the ratio P_e/P_w that appears in equation 3, and if the type I mortality was the same for east and west, the ratio would still be the same as the estimates used to estimate T_w.

4

Fish Stock Assessment

In this chapter, the committee reviews the principal components of the assessment of Atlantic bluefin tuna and performs a reanalysis based on its review and the significant rates of transfer across the Atlantic Ocean presented in Chapter Three. Three major components of assessment are growth, catch per unit effort (CPUE) indices, and virtual population analyses (VPA). In the section for each component, the nature of the review and of the reanalysis performed is described.

GROWTH

It was beyond the committee's purview to undertake a detailed examination of growth. Developments in the analysis of growth curves include the generalization of many models of sigmoid growth into a single function (Richards, 1959; Fletcher, 1975; Schnute, 1981). The Richards (or Schnute case 1) growth model has four parameters and contains many three-parameter growth models as special cases (e.g., von Bertalanffy, Gompertz, logistic). The technique of nonlinear least-squares regression is frequently used to estimate parameters.

Environmental and/or genetic factors may affect the shape parameter of the growth models for Atlantic bluefin tuna (in particular, in the Mediterranean Sea versus the Gulf of Mexico). The shape parameter may be an important aspect of growth that is sensitive to environmental conditions or genetic differences which is most readily adjusted by the organism during early life history. Thus, careful comparisons of this shape parameter may be useful for a better understanding of the selective contributions of the two spawning areas.

STANDARDIZATION OF WESTERN ATLANTIC BLUEFIN TUNA CATCH RATES (CPUES)

Introduction

Several indices based on catch rate as measured by CPUE are used in the VPA of Atlantic bluefin tuna, as described in the International Commission for the Conservation of Atlantic Tunas (ICCAT) report (ICCAT, 1993; table 3 gives the annual values). These indices are of critical importance because their trends are influential in determining the final trend in population parameters from the analysis. The committee wanted to evaluate the uncertainty in these indices and therefore obtained data from three indices, as explained below. The methods used to obtain the indices were examined and as a consequence, the committee performed analyses to reestimate annual values for the indices. In addition, hypothesis of trends in the CPUE data and abundance indices were made, using a robust procedure described below. This trend analysis was performed on the new information as well as the abundance indices from Table 3 of the ICCAT report.

Model Development

The use of CPUE as an index of abundance relies on the direct relationship CPUE = $C/E = qN$, where C is catch, E is effort, q is the catchability coefficient, and N is abundance, biomass, or density. A common problem has been that the value of q may depend on various fishing methods, gear types, and environmental conditions. When annual indices of abundance are desired, the annual trend in CPUEs will reflect the abundance trend only if q is constant over years. However, because fishing techniques and environmental conditions may vary annually, CPUEs need to be standardized so that their trend is independent of these other factors. The annual trend in standardized CPUEs will reflect changes in CPUEs that are not attributable to the standardization factors. Therefore, the annual trend of the standardized CPUEs is attributable to factors not used in the standardization procedure, the most important of which is abundance. Nonlinear relationships between CPUE and abundance were not considered in this report owing to data limitations, but it is true that the presence of nonlinear relationships would induce additional uncertainty about population status and should be considered further.

One popular technique for standardizing catch rates is the use of general linear models. Usually, the CPUEs are transformed so that the annual effects and standardization factors will be multiplicative; for example,

$$\text{CPUE} = \exp(\mu + A + V + X + Z + \varepsilon), \tag{1}$$

which is equivalent to the additive model

$$\ln(\text{CPUE}) = \mu + A + V + X + \ldots + \varepsilon, \qquad (2)$$

where A represents annual effects, V represents effects of the fishing operations, X represents environmental effects, ε represents random error term with mean zero and variance σ^2, and Z represents additional effects on catchability.

A serious problem arises when observations of zero CPUEs, whose logarithms are undefined, appear in the data. The usual solution to this problem is to add a constant:

$$\ln(\text{CPUE}+c) = \mu + A + V + X + \ldots + \varepsilon. \qquad (3)$$

Model 3 is equivalent to multiplicative model 1 with the addition of the constant c. For small values of c relative to the magnitude of the CPUE data, the two models have similar structures. However, if the constant is large, the structure of model 3 could be much different from that of model 1. No longer is CPUE directly proportional to abundance in model 3. Thus, the major problem with using a large constant is that the potential exists for altering the basic multiplicative structure of the model. Indeed, the constant induces a structural change in the form of the model and may create interactions and significant effects among other factors, which would not exist without the constant (or with a small constant) and vice versa.

The models for standardizing Atlantic bluefin tuna CPUEs have used constants such as 1 and $10 \times \max(\text{CPUE})$ (Porch and Scott, 1993). The two resulting trends in standardized CPUEs differed. Furthermore, the transformed data did not meet the normality requirements assumed in testing the significance of the factors in the model, making inferences about the significance of the factors imprecise. There is little support in the statistical literature for choice of a large constant in relation to the magnitude of the data (Zar, 1974; Berry, 1987). In Berry's paper, the examples show small constants in relation to the magnitude of the data, except for one case where a large value was deleted to examine sensitivity. In this case the suggestion is made that the data may not need to be transformed at all, rather than having a large constant added to the observations. Berry's paper had the goal of minimizing skewness and kurtosis, but with Atlantic bluefin tuna the primary goal is to obtain a measure of abundance. These two goals may not be compatible. To resolve some of these problems, other methods have been investigated to standardize Atlantic bluefin tuna CPUEs (Porch and Scott, 1993), with emphasis on solving the problem of zeroes, including the Box-Cox transformation suggested by Berry (1987) and the delta-lognormal distribution (Lo et al., 1992).

No matter what transformation is used, back-transformation to the original CPUE scale requires consideration of bias. In the lognormal case, estimating the predicted values as $\exp(\text{mean}(\ln(\text{CPUE})) + \hat{\sigma}^2/2)$ removes much of the bias. For the constant-added log transformation, see Porch and Scott (1993).

Another complication that arises in standardizing CPUEs with general linear models is weighting. Some fishing trips are longer than others. Therefore, these trips sample more area and should be weighted more. In other words, $\Sigma C/\Sigma E$ is less variable than $1/n(\Sigma C/E)$, suggesting that effort weighting is desirable (see Quinn et al., 1982); that is:

$$\Sigma C/\Sigma E = qN. \tag{4}$$

The underlying model for catch in this situation is:

$$C = \beta E + \sqrt{E} \cdot \varepsilon, \tag{5}$$

where ß is the CPUE parameter (qN), and ε is an error term with mean 0 and constant variance σ^2. Note that this error structure is different from that in models 1 and 2. A recasting of model 5 as:

$$C/\sqrt{E} = \beta \cdot \sqrt{E} + \varepsilon, \tag{6}$$

suggests that the variable $C/\sqrt{E}$ is a candidate for further analysis. Thus, the standardization model 1 or 2 could be recast with the error structure of model 6:

$$\begin{aligned} C/\sqrt{E} &= \sqrt{E} \exp(\mu + A + V + X + \ldots) + \varepsilon \\ &= f(\Theta) + \varepsilon, \end{aligned} \tag{7}$$

where Θ is the set of unknown parameters $\{\mu, \{A\}, \{V\}, \{X\}, \ldots\}$. By using a least-squares criterion, parameters can be estimated by minimizing:

$$SSQ = \Sigma(C/\sqrt{E} - f(\Theta))^2 = \Sigma E \ (C/E - f(\Theta)/\sqrt{E})^2, \tag{8}$$

which shows that this is an effort-weighted nonlinear least-squares procedure. For this model, predicted CPUEs do not require any bias adjustment. Subsequent to the writing of the first draft of this report, the committee found a similar method that is available in the commercial software package S-Plus®.[1] Application of the S-Plus® log-linear model with Poisson error assumption produces results similar to the ones given in this report.

---

[1] Statsci, 1700 Westlake Ave. N., Suite 500, Seattle, WA 98109.

Analyses

The committee was able to obtain data for three indices: the rod and reel survey for small fish (ages 1 to 5), the rod and reel survey for giant fish (ages 8+), and captains' logbook data for giant fish. The committee reanalyzed these data as explained below.[2]

Rod and Reel Indices for Small Fish

Catch rates of small bluefin tuna were standardized by the general linear model method of Brown and Browder (1993), except that the constant for the transformation, ln(CPUE+c), was set to approximately 0.3, which is 0.1 × mean CPUE. This value was small enough to maintain the character of the original data but large enough to prevent the indeterminacy at the origin, which occurs in the logarithm function. It was assumed that each combination of boat name and state of operation represents a unique boat. Although this is not entirely true because more than one boat may have the same name, the approximation is fair. Then, to test the assumption that the CPUE trend is affected by the experience of skippers, three data sets were made. The first data set included all boats. In the second, trips by boats for which no bluefin tuna had ever been sampled by the National Marine Fisheries Service were excluded. This restricted the sample to boats that had some experience catching bluefin tuna. The third data set had more restrictive rules: only boats for which bluefin tuna had been sampled in more than one year were used. No estimates were made prior to 1984 for data sets two and three, because boat names were not recorded before then. Also, 1984 data were excluded from the small bluefin tuna analysis because boat type data used for standardization were absent in 1984. Back-transformed, corrected, least-squares means (Brown and Browder, 1993) from the general linear models on all three data sets produced similar results, as shown in Table 4-1. The Standing Committee on Research and Statistics (SCRS) analyses originally showed a dramatic decline in 1992; however, after data processing errors were corrected by Inter-American Tropical Tuna Commission scientists under the committee's direction, the decline is much smaller. Effort weighting was attempted for this logarithm model to see if weighting altered the results. However, this weighting caused the method to produce unreliable results for some unknown reason.

Because the choices of the weighting factor and the constant of transformation affect the estimated trend in standardized CPUE, we used the effort-weighted, nonlinear, least-squares regression model (model 7) of the form $C/\sqrt{E}$

---

[2]The committee considered conducting bootstrap analysis, but didn't because of the time constraint and the problem of how to run this analysis without having all the raw data.

TABLE 4-1 Estimates of small bluefin tuna CPUE (catch per 100 hours) from rod and reel and handline.*

Year	All Boats	Trips	Only Boats for Which Tuna Were Sampled	Trips	Only Boats for Which Tuna Were Sampled in Two or More Years	Trips
80	56.1	720	N/A	—	N/A	—
81	17.1	454	N/A	—	N/A	—
82	143.1	308	N/A	—	N/A	—
83	29.6	784	N/A	—	N/A	—
84	N/A	—	N/A	—	N/A	—
85	30.0	416	37.2	383	42.5	209
86	46.0	583	62.1	499	62.5	345
87	67.7	491	84.6	420	96.4	276
88	49.9	357	49.5	307	55.9	195
89	59.5	646	73.4	575	77.7	401
90	39.8	667	42.7	604	45.6	440
91	63.8	619	55.1	561	51.6	345
92	37.8	590	41.0	398	59.2	197

N/A = data not available
*The committee was unable to obtain confidence intervals due to time constraints.

$= \sqrt{E} \times \exp(Y + B + A) + \varepsilon$, where C is the catch, E is the effort, Y represents the annual effects, B represents the boat type effects, A represents the area effects, and ε, is the error. The results (Table 4-2 and Figure 4-1) were similar to those of the general linear models, suggesting no trend in abundance effects over time.

Rod and Reel Indices for Giant Bluefin Tuna

These data were analyzed in a similar fashion as the data for small fish. Both the back-transformed, corrected, least-squares means from the general linear model of $ln(\text{CPUE}+0.05)$ and the effort-weighted nonlinear least-squares regression model were used. The significant factors used were area, fish targeted, and month. Fish targeted and area were combined into three categories, GOMA, STNH[3] targeting giant bluefin tuna and STNH targeting marlin and other tunas. Because the second category was present only in 1983, the effect of category 2 on 1983 is removed. No annual trend was apparent after 1983 (Tables 4-3 and 4-4, Figure 4-2). The SCRS analyses showed a dramatic decline in

---

[3] The acronyms GOMA and STNH stand for the Gulf of Maine area and the southern New England to New Jersey area, repectively (Cramer and Turner, 1993).

TABLE 4-2 Relative abundance of small bluefin tuna from effort-weighted nonlinear least-squares regression (1992=1.00), including lower and upper 95% confidence limits.[a]

Year	All Boats	Lower	Upper	Only Boats for Which Tuna Were Sampled	Lower	Upper	Only Boats for Which Tuna Were Sampled in Two or More Years	Lower	Upper
80	1.93	1.80	2.08	N/A	–	–	N/A	–	–
81	0.43	0.32	0.59	N/A	–	–	N/A	–	–
82	2.12	1.94	2.32	N/A	–	–	N/A	–	–
83	0.77	0.65	0.91	N/A	–	–	N/A	–	–
84	N/A	–	–	N/A	–	–	N/A	–	–
85	0.80	0.62	1.03	0.87	0.67	1.13	0.76	0.55	1.06
86	1.04	0.87	1.24	1.12	0.93	1.33	0.79	0.62	1.00
87	1.63	1.41	1.88	1.85	1.60	2.13	1.43	1.20	1.70
88	1.05	0.83	1.32	1.08	0.86	1.37	0.84	0.63	1.12
89	1.55	1.37	1.74	1.67	1.48	1.87	1.24	1.07	1.43
90	0.78	0.63	0.95	0.80	0.65	0.99	0.55	0.42	0.73
91	1.84	1.67	2.04	1.89	1.70	2.10	1.64	1.46	1.86
92	1.00	0.83	1.21	1.00	0.80	1.24	1.00	0.78	1.28

N/A = data not available

[a]Results from an earlier version of the method. Results from a later version differ slightly from the ones in this table but do not cause any changes to the conclusions of the report.

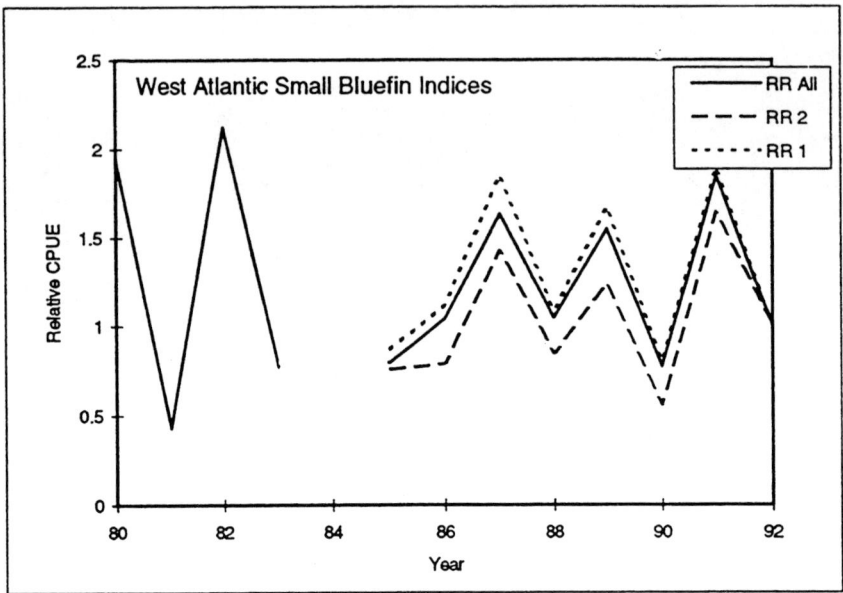

FIGURE 4-1 Western Atlantic Ocean small bluefin tuna indices.

1992; however, after data processing errors were corrected by Inter-American Tropical Tuna Commission scientists under the committee's direction, no such decline could be detected.

Captains' Logbook Data for Giant Bluefin Tuna

Captains' logbook data were summed for each permit number in each year over all days of fishing to reduce the number of zero values. Only data from the general category permits were analyzed, which eliminated the harpoon category data, for which different regulations exist. The average month of fishing, average gear type, and average percentage of spotter plane usage were calculated by weighting observations by the number of days of effort. Average gear type was calculated by assigning gear codes of 1 to harpoon-type gear, 2 to mixed gear, and 3 to nonharpoon-type gear, and then computing a weighted average across all logbook records for a given year/captain/permit number combination.

Three analyses of CPUEs were done from 1988 to 1993. The first analysis uses model 3, a general linear model on ln(CPUE+0.05), weighted by ln(days+1), back-transformed without the correction for the standard error (median estimate). The second analysis is the same as the first but with a $\sigma^2/2$ correction (mean estimate). The third analysis is the effort-weighted, nonlinear, least-squares regression model (model 7). For all analyses, only gear and captain were signifi-

TABLE 4-3 Estimates of giant bluefin tuna CPUE (catch per 100 line-hours) from rod and reel and handline, including number of trips and lower and upper 95% confidence limits.

Year	All Boats	Number of Trips	Lower	Upper	Only Boats for Which Tuna Were Sampled	Number of Trips	Lower	Upper	Only Boats for Which Tuna Were Sampled in Two or More Years	Number of Trips	Lower	Upper
83	4.25	1,589	3.09	5.82	N/A	0	–	–	N/A	0	–	–
84	1.60	1,201	1.04	2.45	1.03	908	0.69	1.83	0.53	222	0.16	1.50
85	1.33	429	0.70	2.48	0.53	357	0.24	1.10	0.42	208	0.12	1.20
86	0.44	122	0.11	1.45	0.49	110	0.10	1.85	0.38	59	0.0004	3.26
87	0.55	1,917	0.37	0.83	0.38	1,275	0.25	0.58	0.37	788	0.20	0.64
88	1.01	760	0.60	1.68	1.09	508	0.59	1.98	1.12	352	0.52	2.37
89	0.85	1,199	0.54	1.32	0.72	746	0.45	1.20	0.47	418	0.21	0.97
90	0.78	1,635	0.48	1.16	0.85	1,067	0.56	1.29	0.86	616	0.47	1.54
91	0.77	1,478	0.51	1.15	0.53	1,050	0.34	0.81	0.32	413	0.13	0.68
92	0.99	1,010	0.62	1.55	0.57	430	0.29	1.10	0.67	251	0.24	1.72

N/A = data not available

TABLE 4-4 Relative abundance of giant bluefin tuna from effort-weighted nonlinear least-squares regression (1992=1.00), including lower and upper 95% confidence limits.

Year	All Boats			Only Boats for Which Tuna Were Sampled			Only Boats for Which Tuna Were Sampled in Two or More Years		
		Lower	Upper		Lower	Upper		Lower	Upper
83	2.39	1.47	3.88	N/A	-	-	N/A	-	-
84	1.32	0.80	2.17	1.28	0.83	1.98	1.18	0.65	2.17
85	1.05	0.60	1.83	0.88	0.55	1.43	1.17	0.67	2.06
86	0.72	0.25	2.11	0.56	0.22	1.43	0.96	0.38	2.52
87	0.62	0.33	1.18	0.58	0.33	0.99	0.72	0.38	1.35
88	1.10	0.62	1.94	1.01	0.62	1.66	1.45	0.85	2.49
89	0.97	0.55	1.71	1.15	0.73	1.83	1.76	1.04	2.96
90	0.76	0.44	1.33	0.86	0.54	1.40	1.41	0.84	2.35
91	0.99	0.59	1.67	0.99	0.63	1.56	1.23	0.65	2.30
92	1.00	0.60	1.71	1.00	0.60	1.67	1.00	0.49	2.04

N/A = data not available

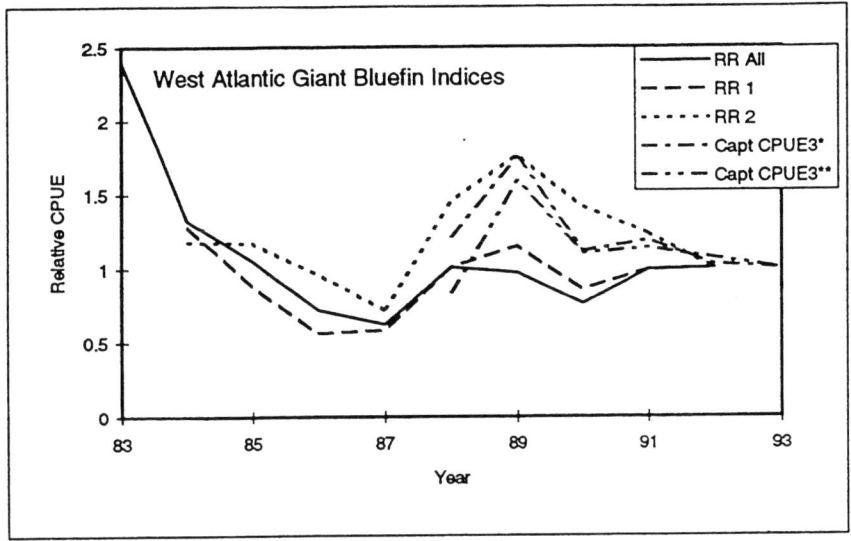

FIGURE 4-2 Western Atlantic Ocean giant bluefin tuna indices.

cant (Table 4-5 and Figure 4-2). Month and plane use were significant before captain and gear were added to the model; therefore, captain and gear contain the information on month and plane use, making the latter two factors redundant. No significant annual trend was observed in the indices produced by any of the analyses.

Comparison of Captains' Logbook Data and Rod and Reel Survey for Giant Bluefin Tuna

Neither the captains' logbook data nor the rod and reel survey data show a trend in the standardized CPUEs of giant bluefin tuna during the years of overlap, 1988 through 1992 (Figure 4-2). Furthermore, annual trends restricted to experienced boats also were not significant. One interesting feature of Figure 4-2 is that the rod and reel index with greater restrictions on selection of data (marked RR 2) has a trend more similar to that of the captains' index, although the trends over time are not significant, as shown below.

One concern in interpretation of CPUE indices is the potential for CPUEs to be overly optimistic regarding trends in abundance due to increasing catchability over years. Catchability can increase if improvements to gear, gains in skipper experience, or advances in technology for finding fish occur. In these results, experience was accounted for by deleting observations from vessels in which bluefin tuna catches were not sampled. Although this measure of experience is imperfect, the results did not differ from this factor. Further understanding of

TABLE 4-5 Estimates of relative CPUE of giant bluefin tuna from captains' logbook data.

Year	ln(CPUE+.05)	STDERR	CPUE1	CPUE2	CPUE2 trips	CPUE3*	CPUE3* trips	CPUE3**	CPUE3** trips
88	0.09	0.28	1.10	1.26	111	1.20	111	0.83	105
89	0.50	0.27	1.72	1.89	148	1.75	148	1.59	148
90	0.18	0.27	1.22	1.38	170	1.11	170	1.10	170
91	0.13	0.26	1.16	1.30	159	1.19	159	1.13	154
92	0.27	0.27	1.35	1.50	129	1.03	129	1.07	129
93	0.00	—	1.00	1.00	85	1.00	93	1.00	93
INT	−0.23	0.59	—	—	—	—	—	—	—

NOTE: The model used was ln(CPUE + 0.05) = μ + year + gear + captain + error. where μ = INT = intercept. Case weights were ln(days + 1)

CPUE1 is back-transformed ln(CPUE + 0.05) without correction for the standard error.
CPUE2 is the same as CPUE1, but corrected for the standard error.
CPUE3 is estimated by effort-weighted nonlinear least-squares regression.

*Contains two additional logbooks for 1993, provided during analysis.
**Eliminating three suspected outliers with l residuals l > 4.9 or l standard residuals l > 2.87.

changes in catchability might require experimental efforts or development of fishery-independent indices.

Trend Analysis

A nonparametric procedure for testing whether there is a trend in a set of data is the Mann-Kendall statistic with Sen's slope estimator with confidence intervals, as described by Gilbert (1987). Advantages of this procedure are (1) it is applicable to serially correlated data, (2) it can accommodate gaps in the data series, and (3) it is a robust procedure with respect to the distribution of the data. A possible disadvantage is that its power may be reduced compared with a parametric procedure for small sample sizes.

First, trend analysis was performed on all indices listed in Table 3 of ICCAT (1993). Some data were truncated to accommodate only the most recent observations, as indicated by the sample sizes shown in the third column in Table 4-6. Only four of the 16 analyses considered for western Atlantic fish were significant ($P \leq 0.05$). The four significant trends had negative slopes; no significant positive slopes were found (Table 4-6).

Second, the sets of CPUE and abundance estimates in Tables 4-1 to 4-5 (rod

TABLE 4-6 Trend analysis of abundance indices for western Atlantic bluefin tuna. Stock estimation abundance indices analyzed were taken from Table 3, ICCAT (1993).

No.	Index	Sample Size	P-value	Slope
1	Larvae	13	0.017*	−0.146
2	Tended Line	10	0.002*	−0.020
3	JLL 81-92	12	0.891	0.003
4	JLL 79-81 10+	8	0.016*,**	−0.234
5	JLL 70-81 8+	12	0.244	0.023
6	HJLL 1-7	12	0.016*	−0.071
7	USGM 8+	6	1.000	−0.010
8	USEEZ 3	6	1.000	−0.134
9	USEEZ 4	6	0.453	−0.075
10	USEEZ 5	6	1.000	−0.014
11	USEEZ 6	6	0.453	0.120
12	USEEZ 7	6	0.453	0.144
13	CUSC 8+	5	1.000	−0.004
14	USC 8+	10	0.074	−0.330
15	USC 6-7	6	1.000	0.092
16	USC 1-5	11	0.276	2.901

*Significant trends—all with negative slopes ($P \leq 0.05$).
**From Gilbert (1987), Appendix A, Table A18 (to provide a more accurate P-value).

and reel adjusted for experience and captains' logbook data) were examined for trends over time with the same procedure. No significant trends were found in any set (Table 4-7).

Recommendations

Quality Control Because most of the methods give similar results, improvements in quality control should improve the assessments more than anything else. For bluefin tuna errors during data collection, entry, and processing should be avoided. All present efforts being done to eliminate invalid data should be continued. Because most of this work is not being documented, the committee cannot determine whether its recommendations below may already have been implemented. If they are being implemented, they need to be documented.

The first link in the chain of quality control is data collection. Errors introduced at this stage may be propagated through data entry, processing, and analysis unless they are identified and corrected. For example, samplers should be carefully supervised and evaluated. Fishermen who tend to give invalid information should continue to be identified and eliminated from the analyses.

The second potential source of error is data entry. The two most common methods of preventing new errors being introduced by data entry are error-checking programs and double entry. Double entry and programming and running error-checking programs are more cost effective than obtaining new data.

The third source of error is in the data processing. Because Atlantic bluefin tuna data come from more than one source in more than one format, this task may not be trivial. In fact, data processing algorithms can be larger and more complicated than the analysis itself. Errors introduced during data processing can be serious. For example, an error in preprocessing the data for Atlantic bluefin tuna resulted in indices that showed a dramatic decline in all sizes of bluefin tuna in the 1992 ICCAT SCRS assessments. When the error was corrected, this apparent decline disappeared or there was no decline. The probability of making these errors can be reduced by detailed documentation of data processing procedures, examining data summaries, double coding, and using software for detecting programming errors. Although documenting data processing is time consuming, trying to figure out what was done years after the analysis can be impossible. Data summaries of the processed data are a useful but limited means of detecting errors. Double coding (having two programmers write the same code), like double data entry, greatly reduces the probability of errors. Only if both programmers make the same error will it go undetected. Compilers have bugs, too, so if there is no additional cost, different compilers (or programming languages) could reduce the number of otherwise undetectable errors. In addition, software exists for checking programs for common programming errors. In fact, without such software, program sizes would be limited by

TABLE 4-7 Trend analysis results for CPUE indices of Tables 1-5.

Source	n	Prob exceeding z	Sen slope
Small bluefin tuna CPUE from rod and reel and handline			
AB[a]	12	0.631	1.067
BS[b]	8	0.711	−1.769
BS2[c]	8	1.000	0.039
Small bluefin tuna, effort-weighted least squares			
AB	12	1.000	0.002
BS	8	0.387	0.035
BS2	8	0.901	0.015
Giant bluefin tuna from rod and reel and handline			
AB	10	0.152	−0.107
BS	9	1.000	−0.015
BS2	9	1.000	0.001
Giant bluefin tuna, effort-weighted least squares			
AB	10	0.211	−0.040
BS	9	0.532	0.018
BS2	9	0.917	0.007
Captains' data, giant bluefin tuna			
ln(CPUE+.05)	6	0.707	−0.051
CPUE1	6	0.707	−0.066
CPUE2	6	0.707	−0.074
CPUE3*	6	0.19[d]	−0.044
CPUE3**	6	0.707	−0.033

[a] AB = all boats.
[b] BS = only boats for which bluefin tuna were sampled.
[c] BS2 = only boats for which bluefin tuna were sampled in 2 or more years
[d] based on table A18 in Gilbert (1987).
*Contains two additional logbooks for 1993, provided during analysis.
**Eliminating three suspected outliers with | residuals | > 4.9 or | standard residuals | > 2.87.

the programmer's error rate. For example, a FORTRAN "lint" is software for checking for possible errors in FORTRAN programs.

Methods Because most of the estimation methods gave similar results, the choice of methods may not be as critical as quality control. However, the committee recommends the effort-weighted, nonlinear, least-squares method (our model 7), because it does not violate the traditional model $C/E = qN$, i.e., it does not depend on the choice of the constant of transformation, and it behaves well with weighting. The biggest improvements to the methods would be to thoroughly document them. Methods must be described in enough detail that another scientist could reproduce the same results.

Another consideration for estimating abundance trends is the regional distribution of effort relative to the geographic distribution of fish abundance. Where there is more than one fishing area, CPUEs in each region are indices of abundance in that region. Weighting observations by the size of their respective fishing areas is one appropriate way of estimating the overall abundance trend (Quinn et al., 1982). Standardizing CPUEs by regional effects will be sufficient when all areas have the same time trend. Another possibility is to consider combining different indices into an overall index. This could involve a single large general linear model analysis of all data sources in which CPUEs are standardized by data source or analysis of the correlations among the different indices.

POPULATION ASSESSMENT

SCRS Base Case

In the management of most fisheries, mathematical models are regularly used to assess the status of an exploited population. These methods provide an integrated framework to assess the consistency of the information available about the population with different hypotheses about its status. Many of these methods belong to the family of VPA techniques. In the core of the VPA, there is a model for the change in the age-specific abundance with time. The predictions provided by the model can then be compared with independent information about the population and a measure of the discrepancy between the model predictions, and the data can be used to tune the parameters of the model through some estimating procedure. There are many different implementations of this fairly general approach, and the differences among them usually reflect the trade-offs between the personal preferences of the scientists and the restrictions imposed by available data.

The methodology used in the SCRS assessments of Atlantic bluefin tuna, ADAPT,[4] falls into this general category. It is based essentially on a back-calculation procedure that allows reconstruction of the abundance of each cohort in the population on the basis of the catches and estimates of natural mortality. An important assumption in this model (as applied by the SCRS working group)

---

[4]ADAPT refers to the name given for a method in: Gavaris, S. 1988. An adaptive framework for the estimation of population size. Can. Atl. Fish. Sci. Advisory Comm. Res. Doc. 88/29. 12p.

is that the population is closed (i.e., no migration is permitted). Four population abundances for different ages in the past year are estimated as parameters in the model, based on a number of independently derived indices of abundance for different groups of ages. ADAPT incorporates several restrictive assumptions in order to minimize the number of parameters being estimated. One of the main assumptions (in the SCRS application to the western Atlantic bluefin tuna) is that the ratio between fishing mortalities of the last two age groups is constant for three blocks of years. The second major assumption of ADAPT has to do with what is known as the separability assumption, common in other integrated methods, which defines the fishing mortality as being the product of age-specific and year-specific components. ADAPT does not explicitly incorporate the separability assumption, although it uses age-specific components (named selectivities or partial recruitments) to estimate abundances for all age classes in the last year. However, these selectivities are estimated by using a separable VPA that uses the same catch data used by ADAPT but with a different assumption about the error structure: the separable VPA assumes that the catches are observed with error, while ADAPT does not. The separable VPA also provides estimates of the ratio between the fishing mortalities of the oldest two age groups.

Another important assumption is that the age composition of the fish caught is known without error. In fact, there is considerable uncertainty about the aging procedure, and variability in the size at age is not incorporated (Butterworth and Punt, 1993). Atlantic bluefin tuna are not aged directly (such as from counts of otolith rings), but instead age is inferred indirectly from measured length of the fish. Thus, ADAPT assumes that there is no error in catch-age information and that all error is contained in the indices. Model parameters are fitted by minimizing a sum of squares function. The sum of squares component may be weighted equally or else the weights may be estimated by the technique of iteratively-reweighted least squares. In SCRS assessments, equal weighting has been used most recently. The committee followed the method of equal weighting of normalized indices as was employed in the 1993 SCRS assessments of Atlantic bluefin tuna. The normalization procedure consists simply of construction of a so-called normalized index by dividing each value of the original index by its overall average. The method of iterative reweighting should be used with some caution. This method favors indices with low inter-annual variability in trend in relation to abundance. As bias cannot be directly observed, an index with a biased trend, but little variablity, can dominate the analysis.

The Role of Indices

Beside the structural assumptions mentioned above, the indices used in the tuning greatly influence the results. In the SCRS assessment, this influence differs among indices, because they are not uniformly related to different age groups. For example, the abundances of younger fish in recent years are entirely determined by the U.S. rod and reel index for small fish, because it is the only one that includes ages 1 and 2. At the other extreme, several indices offer conflicting views of the trend in older age groups: while the Japanese longline index for the northwestern Atlantic Ocean suggests that the population has been stable over the past 18 years, other indices suggest declines. ADAPT reconciles this difference by weight of evidence, which favors the bulk of indices with declines; as a consequence, the Japanese index is the major contributor to the sum of squared residuals, a measure of the discrepancy between the data and the model. To further illustrate this, when we incorporated three data points from the Japanese index that were omitted by mistake from the 1993 SCRS assessment (ICCAT, 1993: data for years 1976, 1983, and 1986), the estimated parameters showed little change but the sum of squares increased.

No indices related to the older age groups are available prior to 1974. The estimated abundances of those ages and years depend almost entirely on the assumed ratios of the fishing mortalities for the last two age groups (ratio of F for age 9 to F for age 10+) in the existing indices.

Is the SCRS Base Case Reasonable?

While some of the structural assumptions incorporated in ADAPT might seem too restrictive in principle, in practice, their consequences are difficult to assess. The assumption of constant ratios in fishing mortalities among age groups is critical to the estimability of cohort abundances, although this constraint can be relaxed to some extent by allowing the ratios to depart from the input values and incorporating a component in the sum of squares that would penalize large departures.

More interesting is the comparison with an assessment with the Stock Synthesis Model (SSM), which was reported in the ICCAT document (Porch et al., 1993). The SSM approach is a member of the family of integrated methods but differs from ADAPT in several ways, being more complex and having wider flexibility in incorporating more sources of information. In particular, SSM explicitly incorporates the separability assumption and allows for variability in the aging of the catches, making explicit predictions about body length distributions in the catches. Therefore, an additional piece of information is being incorporated into the estimating procedure. The interesting feature of this analysis is that, although the ADAPT and SSM are structurally different, they yielded similar results when the contribution of the length distributions in the estimation was downweighted. That would seem to indicate that the basic conclusion obtained with ADAPT regarding the trends in spawning stock biomass is robust to these structural differences. However, when the data provided by the indices and the length distributions are weighted equally, the spawning stock biomass seems to be stable over the past five years (1988-1993), as opposed to the decline indicated by the SCRS base case. This suggests that the procedure used to estimate the catch-at-age matrices for ADAPT (cohort slicing) is not adequate and that inclusion of additional information (such as length distributions) directly into the assessment procedure does not support the recent trends in spawning biomass obtained in the SCRS base case.

Are Uncertainties Adequately Incorporated into the SCRS Assessment?

In any stock assessment, uncertainties limit confidence about the status of the population. These uncertainties can originate either from the data, owing to sampling variability or biases, or from the model, owing to an incorrect specification of the structure of the model. Usually, it is easier to account for uncertainties that originate from the data, because the number of possible models is large.

The assessment carried out by the SCRS on the western component of the Atlantic bluefin tuna incorporated a number of uncertainties arising from the data (see Restrepo et al., 1993). However, uncertainties resulting from the model specification were dealt with inadequately. For example, the most obvious alternative model not considered by the SCRS assessments is the one that assumes that exchange occurs between the eastern and western components of the fisheries.

The Two-Area Mixed-Stock Case

Assessments by the SCRS working group have focused on one of the possible stock hypotheses for the Atlantic bluefin tuna, namely, that there are western and eastern stocks and that no exchange occurs between them. Using the results of the analyses of tagging data, we can explore the consequences of the alternative hypothesis of mixing between east and west. To do that, the basic ADAPT methodology was modified to carry out assessments simultaneously for both areas, with a link between the eastern and western components of the population provided by the following equations:

$$N_{w,a+1,t+1} = N_{w,a,t} \exp[-(M+F_{w,a,t})] (1-T_w) + N_{e,a,t} \exp[-(M+F_{e,a,t})] T_e \tag{9}$$

$$N_{e,a+1,t+1} = N_{e,a,t} \exp[-(M+F_{e,a,t})] (1-T_e) + N_{w,a,t} \exp[-(M+F_{w,a,t})] T_w \tag{10}$$

plus two equations for the plus group (fish of ages 10 or greater) in the areas, where w represents the western area, e represents the eastern area, a is age, t is year, N is abundance, M is natural mortality, F is fishing mortality, and T is the movement rate from one area to the other. Migration was assumed to occur at the end of the year. This approach is essentially the one applied by Butterworth and Punt (1993). Although the movement rates and natural mortality rates are set to fixed constants for each analysis, this is not a static analysis; instead, it accommodates nonequilibrium dynamic trajectories of stock abundance. Changes with time in natural mortality or movement rates would likely require additional experimental data, such as from tagging experiments, or possibly further analysis of existing data. The transfer rates in (9) and (10) are annual rates, whereas the rates estimated in Chapter 3 are instantaneous. However, because the rates are small, they should be quite similar.

The analysis was based on data for the period 1970 to 1992. Catch at age for the eastern component was not available for 1992, so it was assumed that the catches in that year were the same as in 1991. Movement rates assumed for the population were derived from the analysis of tagging data: 1% per year from west to east for all ages, and 2 and 3% per year from east to west for ages 6 and younger. The indices used for the tuning were the same indices used in the last SCRS assessments, with the exception of the U.S. rod and reel indices for small and large fish, which were revised for this assessment. The index derived from the captains' logbook data was not used because it was not completely independent of the U.S. rod and reel index for large fish and because it showed essentially the same trend as the latter. An additional inconsistency in the procedure applied to this analysis is that the selectivities by age used for both components of the population were estimated from results of the separable VPA, assuming isolated populations in the east and west.

Table 4-8 shows the ratios of spawning stock (fish of ages 8 and older) abundances (N_{1993}/N_{1975}; N_{1993}/N_{1988}) and biomasses (B_{1993}/B_{1975}; B_{1993}/B_{1988}) from different analyses. These analyses (cases), described below, have different levels of exchange and other variations to assess the effects of this variability on the results. Tables 4-9 and 4-10 list the vectors of exploitation rates and fishing mortalities by age in 1992. The contribution of each index to the total sum of squared residuals (a measure of the discrepancy between the predictions of the model and the data) is given in Table 4-11.

Results indicate that as we allow for movement of fish between areas, it becomes difficult for the method to reconcile the trends suggested by the indices and the predictions of the model. The assessment tries to accommodate the large number of fish coming from the east. This is a consequence of the large differences in abundance of bluefin tuna on the eastern compared to the western side of the Atlantic Ocean: a contribution to the west of 2% (**case 1**, our base case) or 3% (**case 2**) of the abundance in the eastern Atlantic Ocean represents an important proportion of the western component. In years when either the recruitment in the east is high (e.g., 1983) or the indices in the west show a declining trend, the model tries to adjust to the large numbers of fish moving from east to west by estimating recruitment failures in the west for several years.

Allowing exchange from the east results in a more optimistic appraisal of the status of the western component. Assuming annual exchange rates of 3%

TABLE 4-8 Spawning stock abundance (N8+) and biomass (SSB) ratios (1993/1988 and 1993/1975) for the Western component of the Atlantic bluefin tuna.

Case	Emigration rate		N(8+)		SSB	
	East	West	93/88	93/75	93/88	93/75
1	2%/yr	1%/yr	92.30%	18.10%	99.95%	24.54%
2	3%/yr	1%/yr	126.70%	36.10%	126.29%	45.70%
3	2%/yr	1%/yr	100.30%	21.30%	102.76%	27.35%
4	2%/yr	1%/yr	88.10%	20.60%	98.31%	25.28%
5	2%/yr	1%/yr	92.40%	18.20%	100.32%	24.65%
6	2%/yr	1%/yr	98.90%	20.70%	102.89%	26.98%
7	0%/yr	0%/yr	75.80%	13.80%	87.91%	19.09%
8	2%/yr	1%/yr	129.80%	43.00%	125.87%	50.85%

Note: Case 1: Base case, Case 2: Increase eastern emigration rate, Case 3: Omit GOM index, Case 4: Vary natural mortality at age, Case 5: Delete Canadian CPUE index, Case 6: Increase catch of age 1 in Eastern Atlantic Ocean, Case 7: No migration (SCRS base case with data processing errors corrected), Case 8: Continue migration past age 6.

from east to west and 1% from west to east (case 2), the indication is that the spawning stock in the western Atlantic Ocean has increased to 127% of the 1988 level, but is only 36% of the 1975 level (Table 4-8). A more conservative movement rate from east to west (2% annually, case 1) results in a spawning stock abundance about the same as the 1988 level (at about 92%), but is only 18% of the 1975 level. The predicted and observed values for the different indices for this particular case are shown in Figure 4-3. The overall fit to the indices is good, although there are discrepancies because the directions of trends differ among some indices.

A number of sensitivity runs were made to assess the consequences of changing certain assumptions. One question that arises when we allow for mixing of the two stocks is related to spawning site fidelity. We cannot ascertain at this time whether fish spawn only in the areas where they were born or whether they exhibit a more opportunistic reproductive strategy, spawning wherever the environmental conditions are appropriate, regardless of where they are located. If the former hypothesis is correct, we cannot assume the indices derived from data in spawning areas of the west adequately represent the abundance of ages 8 and older. That is because these age classes would include a number of fish that are not present in the spawning areas. Therefore, a run was done in which the indices corresponding to the Gulf of Mexico were omitted (**case 3**). The results indicate virtually no change in spawning stock abundance since 1988, and a level representing 20% of the 1975 reported abundance.

A separate analysis (also assuming annual movement rates of 2% east-west and 1% west-east) was done allowing for changes in age-specific natural mortal-

TABLE 4-9 Estimated exploitation rates by age in 1992 for the different cases.

	Case 1	Case 2	Case 3	Case 4	Case 5	Case 6	Case 7	Case 8
West								
Age 1	0.013	0.008	0.013	0.016	0.013	0.011	0.015	0.013
Age 2	0.067	0.042	0.066	0.081	0.067	0.060	0.074	0.069
Age 3	0.077	0.048	0.076	0.093	0.077	0.068	0.085	0.080
Age 4	0.041	0.022	0.040	0.052	0.040	0.038	0.044	0.038
Age 5	0.041	0.022	0.040	0.052	0.040	0.038	0.044	0.038
Age 6	0.058	0.045	0.066	0.069	0.057	0.062	0.062	0.077
Age 7	0.058	0.045	0.066	0.069	0.057	0.062	0.062	0.077
Age 8	0.112	0.046	0.093	0.159	0.112	0.096	0.175	0.055
Age 9	0.126	0.053	0.105	0.179	0.126	0.108	0.196	0.063
Age 10+	0.108	0.045	0.090	0.154	0.108	0.093	0.170	0.054
East								
Age 1	0.510	0.672	0.498	0.548	0.515	0.539	0.352	0.534
Age 2	0.584	0.743	0.571	0.622	0.588	0.613	0.414	0.608
Age 3	0.531	0.693	0.519	0.569	0.535	0.560	0.369	0.555
Age 4	0.683	0.805	0.653	0.615	0.639	0.984	0.464	0.503
Age 5	0.624	0.754	0.593	0.556	0.579	0.980	0.410	0.447
Age 6	0.376	0.495	0.351	0.322	0.340	0.932	0.223	0.247
Age 7	0.422	0.548	0.396	0.365	0.383	0.950	0.254	0.281
Age 8	0.176	0.211	0.120	0.289	0.175	0.187	0.198	0.077
Age 9	0.232	0.277	0.161	0.374	0.231	0.247	0.261	0.104
Age 10+	0.611	0.683	0.468	0.804	0.608	0.636	0.658	0.327

ity rate: 0.5, 0.4, 0.3, and 0.2 for ages 1 through 4, and 0.1 for ages 5 and older (**case 4**). As expected, this assessment is closer to the results assuming a low exchange rate, because now the number of fish that move between east and west is much lower. The 1993 spawning stock[5] abundance in numbers is estimated to be 88% relative to the 1988 level and 20% relative to the 1975 level. One caveat to this analysis is that the movement rates were estimated under an assumption that natural mortality was 0.14 for all ages; higher values of natural mortality in the younger ages cause higher estimates of movement rates.

The consequences of deleting the Canadian tended-line CPUE from the set of indices (**case 5**) also were analyzed. This resulted in the same trends for the spawning stock abundance as indicated in the base case with 2% and 1% exchange rates (92% in 1993 relative to 1988, 18% relative to 1975).

The effect of underreporting catches of young fish in the eastern Atlantic Ocean was assessed by increasing the catch of one-year-old fish in the eastern

---

[5]In this section of the report, spawning stock is defined as fish on a given fishing ground (either the east Atlantic Ocean or the west Atlantic Ocean) that are capable of reproduction, but whose spawn localities are unknown.

FISH STOCK ASSESSMENT

TABLE 4-10 Instantaneous fishing mortality rates by age in 1992 as estimated for the different cases considered in the VPA.

	Case 1	Case 2	Case 3	Case 4	Case 5	Case 6	Case 7	Case 8
West								
Age 1	0.014	0.009	0.014	0.017	0.014	0.012	0.016	0.014
Age 2	0.074	0.046	0.073	0.091	0.075	0.066	0.082	0.077
Age 3	0.086	0.053	0.085	0.105	0.086	0.076	0.095	0.089
Age 4	0.045	0.024	0.044	0.057	0.044	0.041	0.048	0.041
Age 5	0.045	0.024	0.044	0.057	0.044	0.041	0.048	0.041
Age 6	0.064	0.049	0.073	0.077	0.063	0.069	0.069	0.086
Age 7	0.064	0.049	0.073	0.077	0.063	0.069	0.069	0.086
Age 8	0.128	0.051	0.105	0.186	0.128	0.108	0.207	0.061
Age 9	0.145	0.058	0.119	0.212	0.145	0.123	0.235	0.070
Age 10+	0.123	0.049	0.101	0.180	0.123	0.105	0.200	0.059
East								
Age 1	0.781	1.238	0.753	0.870	0.791	0.849	0.470	0.837
Age 2	0.964	1.528	0.930	1.074	0.977	1.048	0.581	1.033
Age 3	0.829	1.314	0.800	0.924	0.840	0.901	0.499	0.888
Age 4	1.279	1.868	1.173	1.054	1.126	8.469	0.679	0.764
Age 5	1.080	1.577	0.991	0.890	0.950	7.151	0.573	0.645
Age 6	0.512	0.747	0.469	0.421	0.450	3.388	0.272	0.306
Age 7	0.597	0.872	0.547	0.492	0.525	3.952	0.317	0.357
Age 8	0.208	0.255	0.138	0.368	0.207	0.223	0.238	0.086
Age 9	0.285	0.351	0.189	0.508	0.283	0.306	0.326	0.118
Age 10+	1.042	1.279	0.689	1.856	1.034	1.117	1.190	0.429

side by 50% (**case 6**). Results indicate, once more, that the spawning stock abundance is at virtually the same level as in 1988 (99%) and at a 20% level relative to 1975.

Case 7 is the run assuming isolation of the two components of the population and as such, it represents the SCRS base case with data processing errors corrected. The methodology is essentially the same, but there are differences in the indices: the U.S. rod and reel indices for small and large fish have been revised, and the missing points in the Japanese longline index for the northwestern Atlantic Ocean have been added. In terms of the trends in spawning stock abundance, the assessment in case 7 offers about the same view of the situation of the stock (spawning abundance at 76% relative to 1988 and at 14% relative to 1975) with a better overall fit of the indices, compared to the 1993 SCRS assessment (spawning abundance at 78% relative to 1988 and at 15% relative to 1975). A comparison of the long-term trends in spawning stock biomass between cases 1, 2, and 7 is shown in Figure 4-4.

Because of insufficient tagging of large bluefin tuna in the eastern Atlantic Ocean, the committee was not able to estimate movement rates for those fish, and we assumed a conservative zero probability of movement past age 6. A run

TABLE 4-11 Contribution of each index and total weighted sum of squared residuals for each of the cases considered in the VPA.

	Case 1	Case 2	Case 3	Case 4	Case 5	Case 6	Case 7	Case 8
West								
Larval	0.544	0.723	—	0.533	0.591	0.567	0.505	0.770
JLL Small	0.295	0.508	0.398	0.269	0.322	0.313	0.263	0.436
JLL Medium	0.128	0.202	0.202	0.116	0.138	0.146	0.111	0.206
USRR Small	0.272	0.324	0.337	0.244	0.296	0.283	0.216	0.244
USRR Large	0.060	0.104	0.087	0.062	0.065	0.065	0.057	0.123
Can TL	0.076	0.129	0.113	0.082	—	0.084	0.063	0.161
JLL GoM	0.099	0.086	—	0.092	0.108	0.097	0.101	0.085
JLL NW-Atl	1.611	1.340	2.048	1.610	1.751	1.567	1.699	1.302
USLL GoM	0.141	0.173	—	0.137	0.154	0.148	0.127	0.175
East								
JLL	0.075	0.076	0.096	0.079	0.081	0.075	0.076	0.073
FR PS 2	0.047	0.051	0.070	0.045	0.051	0.049	0.037	0.068
FR PS 3	0.047	0.045	0.074	0.040	0.051	0.048	0.044	0.075
ES Trap 71-81	0.121	0.121	0.159	0.116	0.132	0.121	0.126	0.126
ES Trap 82-91	0.040	0.041	0.053	0.042	0.043	0.040	0.041	0.049
ES BB 70-77	0.036	0.036	0.048	0.037	0.039	0.036	0.037	0.035
ES BB 78-91	0.058	0.063	0.074	0.061	0.063	0.060	0.057	0.058
Total	3.650	4.022	3.757	3.563	3.885	3.698	3.558	3.986

FISH STOCK ASSESSMENT

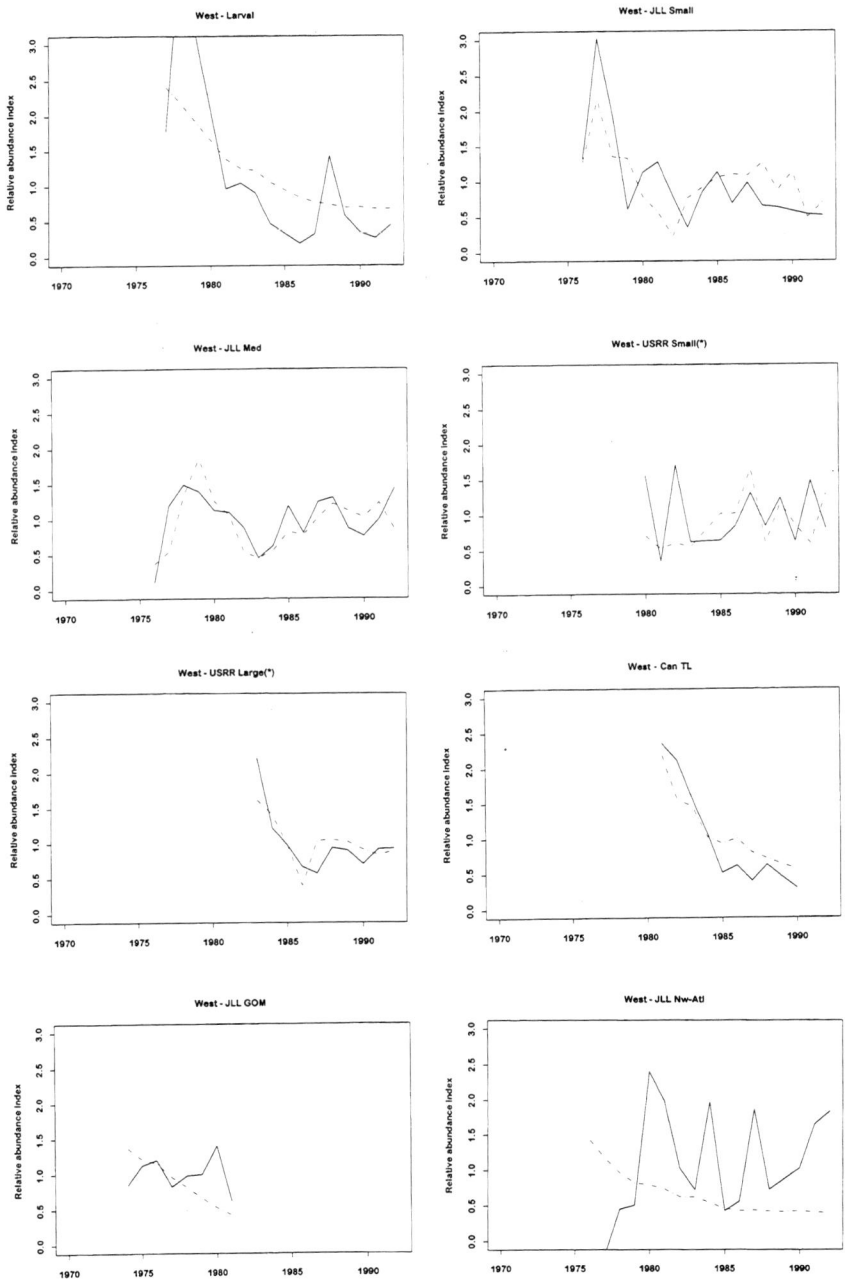

FIGURE 4-3 Observed (solid line) and predicted (dashed line) values of indices for case 1.

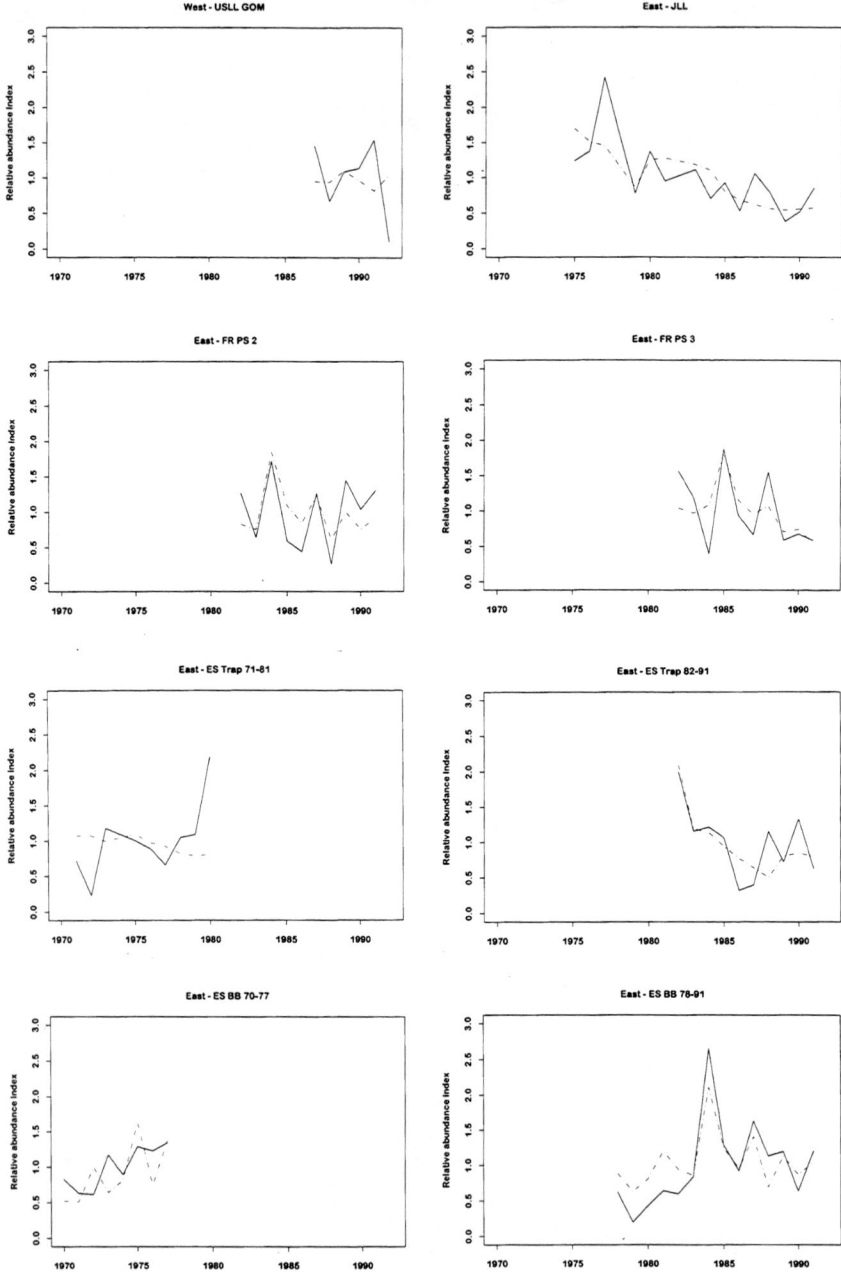

FISH STOCK ASSESSMENT

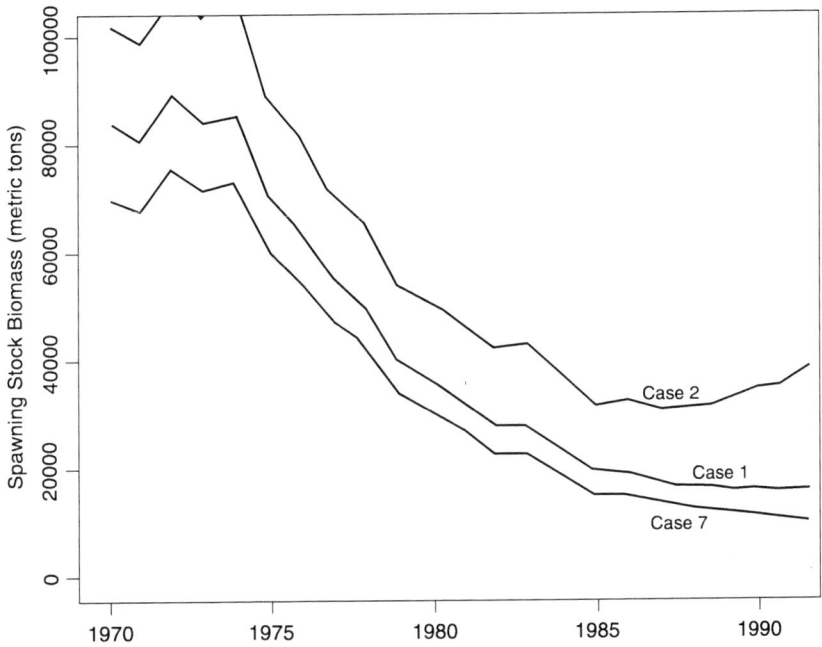

FIGURE 4-4 Comparison of the long-term trends in spawning stock biomass between cases 1, 2, and 7 (see Table 4-8 for case definitions).

assuming that movement rates continue at the 2% annual level for ages beyond age 6 is shown as **case 8**. Trends in spawning stock abundance offer a more optimistic view (spawning stock abundance is 130% relative to 1988 and 43% relative to 1975).

Overall, based on the anaylsis, estimates of fishing mortality for 1992 are much higher in the east than in the west (Table 4-10). The inclusion of movement exacerbates the difference in fishing mortality between east and west. The choice of indices, the change in natural mortality, and possible misreporting tend to be less important than transatlantic movement, except that altering natural mortality (case 4) had a large effect on older ages in the east and misreporting (case 6) caused a large increase for medium ages in the east.

In terms of the fit to the indices, the assessment tends to favor exchange rates that are lower than the ones estimated from the analysis of the tagging data (Table 4-11). This conclusion is consistent with the results described by Butterworth and Punt (1993). As with estimating natural mortality, M, there is probably too little information in the indices to be able to estimate migration; hence, the sum of squares may not be a useful tool for comparison of the different cases. In practice, it is necessary to obtain natural mortality and migration rates from other independent sources and analyses, as was done in this study.

DISCUSSION

Current abundance of Atlantic bluefin tuna in the western Atlantic Ocean has been stable since 1988 based on the analyses above. That result is supported by the U.S. rod and reel indices of both small and giant bluefin tuna; by the U.S. captains' logbook data; and by Japanese longline indices for small, medium, and giant bluefin tuna. Indices that support a decline in abundance since 1988 are those based on data from the Gulf of Mexico (the larval index to some extent depending on the year of reference, but primarily the U.S. longline index and its 1992 data point). Support for an increase in abundance, particularly for the 8+ age group in the west, comes from acknowledging substantial immigration from the eastern Atlantic Ocean; however, that support could be undermined if migration rates are lower in the most recent years from the east; this would not have been detected from the analysis of the tagging data. Assessments also show that the absolute levels of abundance of the 8+ age stock in the western Atlantic Ocean are roughly two to five times greater than the levels estimated in the SCRS report.

Trends of abundance in the western Atlantic Ocean between bluefin tuna in the Gulf of Mexico and those in the fisheries of the western Atlantic Ocean may diverge in the future. This possibility suggests that either different components of the Atlantic stock are being measured or there are errors in the indices. One hypothesis to account for the possible divergence is that the 8+ age group measured by western Atlantic fisheries is not representative of the spawners in the Gulf of Mexico, perhaps because the ocean fish are a mixture containing eastern Atlantic progeny, which exhibit spawning site fidelity and hence do not spawn in the west. Under this hypothesis, control of levels of spawning abundance in the Gulf of Mexico may not be easily accomplished by controls on fisheries in the western Atlantic Ocean: spawning site fidelity may occur with western Atlantic progeny, which may cause migrants in the eastern Atlantic Ocean to represent a portion of the Gulf of Mexico spawners. A second hypothesis to account for the divergence is that bluefin tuna are opportunistic spawners that will spawn whenever environmental conditions are favorable. Under this hypothesis, the environment in the Gulf of Mexico is variable and this variability causes spawners to represent a highly variable component of the 8+ age stock.

An important assessment issue is how to predict future population sizes. Considerations such as definition of spawning stock, partitioning of recruitment between/among areas, and environmental factors could all be critical components of such work. Equally important in such predictions is the recognition that catch quotas have not been constant over the past several decades in the west but rather have been based on estimated changes in stock abundance. Adaptive management techniques can be applied to the projections. As a starting point, constant fishing mortality or harvest rate scenarios can be considered in lieu of constant catch policies as is currently the practice.

Conclusions—Standardization

Estimates of abundance are unchanged, on average, from 1988 to 1993 based on reanalyses of the three indices (two rod and reel indices and one captains' logbook index). The two indices on giant bluefin tuna (one rod and reel index and the captains' logbook index) are concordant. Furthermore, the selection criteria on classes of boats did not alter these conclusions, based on progressively more restrictive data exclusion to remove some inexperienced fishers.

Transformations of catch rates according to logarithm of catch rate plus a large constant are inappropriate because they lack a biological foundation and contradict assumptions about independence of factors on catch rates.

Using nonparametric tests, the Mann-Kendall S-test and Sen's slope estimator for the published SCRS indices, 12 indices were found to have no significant (slope) trends and four indices (see Table 4-6) had significant trends (all negative slopes) at ($P \leq 0.5$). These tests also show that there are no significant trends in any of the CPUE data in Tables 4-1 to 4-5.

Recommendations—Standardization

A data management system should be established for catch indices that includes better documentation, quality control, central archiving, and error checking.

A thorough, in-depth review of all indices from all areas of the Atlantic Ocean, Mediterranean Sea, and Gulf of Mexico is needed.

Alternative methods for estimating stock abundance should be evaluated, including aerial survey, spawner surveys in the Florida Straits, purse seining for young bluefin tuna, and expanded larval fish surveys in the Gulf of Mexico.

Selection criteria should be developed to create a rod and reel index for giant bluefin tuna, which excludes fishermen present in the captains' logbook index. That process would create two independent catch rate indices (the captains' logbook index and the modified rod and reel index on giant fish) that could both be used as tuning indices.

Conclusions—Population Assessment

The estimated spawning abundance of western Atlantic bluefin tuna has declined substantially (to about 20% of that in 1975) since the 1970's. The reasons for this decline are unknown, because the spawner-recruit relationship is uncertain. The current abundance of Atlantic bluefin tuna in the western Atlantic Ocean has been stable since 1988, in contrast to the roughly 50% decline in the age 8+ abundance reported in the 1993 ICCAT report. Reasons for the committee's view of current abundance result principally from two changes made to the SCRS assessments: (1) reanalysis and correction of some accidental errors in the calculation of the U.S. rod and reel index of abundance of bluefin tuna and (2) rejection of the two-stock hypothesis and subsequent reanalysis employing a two-area mixing model.

Given the results of reanalysis, further reductions in catch quotas in the western Atlantic Ocean from 1992-1993 levels cannot be based on a conclusion of a decline in western Atlantic stock abundance since 1988.

The mixing model points to the importance of bluefin tuna in the eastern Atlantic Ocean and Mediterranean Sea when evaluating the status and future of bluefin tuna in the western Atlantic Ocean. Even with relatively low mixing rates, the extremely large abundance of young tunas estimated recently in the eastern Atlantic Ocean causes the abundance of fishable age classes in the west to be strongly influenced by westward migrants.

Fishing mortality in the east is perhaps four times greater than fishing mortality in the west.

Considerable uncertainty exists about the relationship between fish on the fishing grounds and those on the spawning grounds.

Recommendations—Population Assessment

The committee recommends that:

Given the revised view of stock structure, management strategies for the east and west should be revisited and should include minimum size limits, catch limits, and (spawning ground) area closures.

Constant fishing mortality (that is, fixed at a constant annual rate for several years) may offer greater opportunity for rebuilding western stocks or reaching high long-term yields than constant catch quotas over many years.

Future assessments of Atlantic bluefin tuna should be conducted with an assessment procedure that explicitly accounts for mixing. As a consequence, assessments of both eastern and western fisheries should be made together at each SCRS meeting. Given the sensitivity of results to the transfer rates, further inspection of the tagging data is warranted and further analysis to estimate rates should be undertaken.

Alternative models to VPA, such as migratory catch age analysis and migratory stock synthesis analysis, should be considered.

5

Conclusions and Recommendations

GENERAL RECOMMENDATIONS

In response to its charge, the committee makes the following recommendations:

1. Available biological evidence of stock structure although sparse is consistent with a single stock hypothesis for bluefin tuna in the North Atlantic Ocean, with at least two spawning areas. Furthermore, the committee's reevaluation of tagging results confirms that movement of bluefin tuna between the western and eastern Atlantic Ocean is sufficient to alter the previous SCRS stock assessments. **The committee recommends that NOAA/NMFS conduct new scientific assessments explicitly to include mixing of Atlantic bluefin tuna between eastern and western fishing grounds.**

2. In response to the first question posed to the committee by NOAA, the committee concludes that recent ICCAT Standing Committee on Research and Statistics (SCRS) assessments of abundance of eastern and western Atlantic bluefin tuna do not provide the most defensible interpretations of available scientific data. The committee's reanalyses show that there is no evidence that abundance of western Atlantic bluefin tuna has changed significantly between 1988 and 1992. **The committee recommends that NOAA/NMFS use alternative methods of data management, data analyses, and peer review for estimating abundance indices, movement rates, and mixed population assessments (as discussed in Chapters 3 and 4 of this report).**

3. The committee notes that the ICCAT SCRS uses a variety of uncertainty

analyses. **The committee recommends that NOAA/NMFS and ICCAT SCRS act to include transatlantic movement of fish and adaptive management techniques in future uncertainty analyses.**

4. The committee cannot determine the maximum sustainable yield (MSY) for Atlantic bluefin tuna under a one-stock hypothesis with two spawning grounds. Available biological information on stock structure, mixing on the spawning and fishing grounds, spawning site fidelity, and spawner/recruit relationships is too sparse. We do know that the present abundance of bluefin tuna in the western Atlantic Ocean is lower than that in the early 1970s although the committee did not analyze similar data for the bluefin tuna in the eastern Atlantic Ocean. We also know that the present abundance and fishing mortality are much higher in the eastern Atlantic Ocean than in the west, and that some physical mixing occurs between the fishing grounds in the eastern and western Atlantic Ocean. **The committee recommends that NOAA/NMFS reevaluate MSY for Atlantic bluefin tuna.**

RESEARCH RECOMMENDATIONS

The committee notes that research on the biology of Atlantic bluefin tuna is not continuing at an intensity necessary to answer major biological questions pertaining to the management of the fisheries. Therefore, the committee recommends that NOAA/NMFS carry out the research described below using the best available science and techniques within and outside NOAA. For example, research supported by other U.S. government agencies, including the National Science Foundation, the Department of Energy, and the Office of Naval Research, could contribute to the goals of the studies funded by NOAA. Finally, the committee urges NOAA/NMFS to work cooperatively with ICCAT to implement these research recommendations. The following recommendations are not listed in order of importance or priority.

1. Tagging data show that there is movement of bluefin tuna between the eastern and western Atlantic fishing grounds, but the degree of gene flow between spawning areas is not known. Such knowledge is essential in defining population genetic structure and useful for refining stock assessments. **The committee recommends that the one-stock hypothesis be tested rigorously, using the most appropriate technologies capable of detecting contemporary population genetic structure.**

2. Estimates of spawning fidelity to a particular area are essential for stock assessments. **The committee recommends that microconstituent analysis and archival tags be used to provide information on spawning fidelity.**

3. Stock assessments can be refined by better estimates of life history characteristics such as spawning biomass, larval abundance, sex ratio, age at maturity, fecundity, and recruitment. **The committee recommends that spawning**

biomass, sex ratio, age at maturity, and fecundity in the spawning grounds be estimated and that larval performance, as affected by environmental conditions, be studied.

4. The committee recognizes that knowledge of movement patterns is essential for estimating abundance and distribution and that movement rates and patterns may change over time. **The committee recommends that a tagging program be undertaken, with an appropriate combination of conventional, PIT, acoustic, and archival tags to provide improved estimates of the magnitude and patterns of movement. This program should be designed to answer scientific questions pertinent to stock assessment and should be coordinated among all nations involved in the bluefin tuna fishery.**

5. Estimates of abundance are confounded by the interaction between fishing and changes in distribution caused by interdecadal climatic and oceanic variability. **The committee recommends a synthesizing analysis of existing data on distributions of bluefin tuna in relation to spatial and temporal dynamics of major oceanographic features.**

6. The committee notes that a greater use of peer review would have improved the quality of some of the research reviewed during the preparation of this report. **The committee recommends that review of all research proposals and resulting manuscripts include a process of external peer review.**

References

Baglin, R.E. 1982. Reproductive biology of western Atlantic bluefin tuna. Fish. Bull. 80(1):121-134.

Baglin, R.E., M.I. Farber, W.H. Lenarz, and J.M. Mason. 1980. Estimates of shedding rates of two types of dart tags from northwestern Atlantic bluefin tuna (*Thunnus thynnus*). ICCAT Coll. Vol. Sci. Pap. IX:453-462 (SCRS/79/84).

Barber, R.T., and R.L. Smith. 1981. Coastal upwelling ecosystems. In: Analysis of Marine Ecosystems (ed. A.R. Longhurst). Academic Press, New York. pp.31-68.

Bartlett, S.E., and W.S. Davidson. 1991. Idntification of *Thunnus* tuna species by their polymerase chain reaction and direct sequence analysis of their mitochondrial cytochrome *b* genes. Can. J. Fish. Aquat. Sci. 48:309-317.

Baumgartner, T.R., A. Soutar, and V. Ferreira-Bartrina. 1992. Reconstruction of the history of Pacific sardine and the northern anchovy populations over the past two millennia from sediments of the Santa Barbara Basin, California. Calif. Coop. Occanic Fish. Invest. Rep. 33:24-40.

Bayliff, W.H., and L.M. Mobrand. 1972. Estimates of the rates of shedding of dart tags from yelllowfin tuna. Inter-Am. Trop. Tuna Comm. Bull. 15(5):439-462.

Berry, D.A. 1987. Logarithmic transformations in ANOVA. Biometrics 43:439-456.

Bigelow, H.B., and W.C. Schroeder. 1953. Fishes of the Gulf of Maine. Fish. Bull. Fish and Wildlife Service 53:338-347.

Block, B.A., J.R. Finnerty, A.F.R. Stewart, and J. Kidd. 1993. Evolution of endothermy in fish: Mapping physiological traits on a molecular phylogeny. Science 260:210-214.

Brown, C.A., and J.A. Browder. 1993. Standardized catch rates of small bluefin tuna in the Virginia-Rhode Island (U.S.) rod and reel fishery. ICCAT Working Document SCRS/93/67.

Brunenmeister, S. 1980. A summary and discussion of technical information pertaining to the geographical discreteness of Atlantic bluefin tuna resources. ICCAT Coll. Vol. Sci. Pap. IX:506-527 (SCRS/79/95).

Butterworth, D.S., and A.E. Punt. 1993. The robustness of estimates of stock status for the western North Atlantic bluefin tuna population to violations of the assumptions underlying the associated models. ICCAT Working Document SCRS/93/68.

Calaprice, J.R. 1986. Chemical variability and stock variation in northern Atlantic bluefin tuna. ICCAT Coll. Vol. Sci. Pap. XXIV(2):222-254 (SCRS/85/36).

Calaprice, J.R., H.M. McSheffrey, and L.A. Lapi. 1971. Radioisotope x-ray fluorescence spectrometry in aquatic biology: A review. J. of the Fisheries Res. Board Can. 28:1583-1594.

Chapman, D.G., B.D. Fink, and E.B. Bennet. 1965. A method for estimating the rate of shedding of tags from yellowfin tuna. Inter-Am. Trop. Tuna Comm. Bull. 10(5):333-352.

Chapman, D.G., and D.S. Robson. 1960. The analysis of a catch curve. Biometrics 16:354-368.

Chow, S., and S. Inoue. 1993. Intra- and interspecific restriction fragment length polymorphism in mitochondrial genes of *Thunnus* tuna species. Nat. Res. Inst. Far Seas Fisheries Bull. 30:207-225.

Clay, D. ed. 1990. Atlantic bluefin tuna (*Thunnus thynnus thynnus* (L.)): A review. World Bluefin Meeting, May 25-31, LaJolla, Calif.

Clay, D., and T. Hurlbut. 1990. Bluefin tuna (*Thunnus thynnus* L.) fisheries in Atlantic Canada: An historical review and hypothesis of minimum assemblage size. ICCAT Coll. Vol. Sci. Pap. XXXII(2):270-279 (SCRS/89/89).

CLIMAP. 1976. The surface of the ice-age earth. Science 191:1131-1136.

Cole, J.S. 1980. Synopsis of biological data on the yellowfin tuna, *Thunnus albacares* (Bonnaterre, 1788), in the Pacific Ocean. In: Synopses of Biological Data on Eight Species of Scombrids. Special Report No. 2. (ed. W.H. Bayliff). Inter-Amer. Trop. Tuna Comm.: 77.

Collette, B.B., T. Potthoff, W.J. Richards, S. Ueyanagi, J.L. Russo, and Y. Nishikawa. 1984. Scombroidei: Development and relationships. In: Ontogeny and Systematics of Fishes. Special Publication No.1. (eds. Moser et al.). Maryland: American Society of Icthyologists and Herpetologists. pp.591-619.

Cort, J.L., and J.M. de la Serna. 1993. Revisión de los datos de marcado/recaputra de atún rojo (*Thunnus thynnus, L.*) en el Atlántico Este y el Mediterráneo. ICCAT Working Document. SCRS/93/81.

Cort, J.J., and B. Liorzou. 1990a. Larval biology—Eastern Atlantic and Mediterranean. In: World Bluefin Meeting, May 25-31, 1990, (ed. D. Clay). LaJolla, Calif. p. 95.

Cort, J.J., and B. Liorzou. 1990b. Reproduction—Eastern Atlantic and Mediterranean. In: World Bluefin Meeting, May 25-31, 1990, (ed. D. Clay). La Jolla, Calif. pp. 99-101.

Cort, J. J. and B. Liorzou. 1990c. Tagging interpretation—Eastern Atlantic and Mediterranean. In: World Bluefin Meeting, May 25-31, 1990, (ed. D. Clay). La Jolla, Calif. pp. 110-127.

Cramer, J., and S. Turner. 1993. Large bluefin tuna, *Thunnus thynnus*, indices of abundance for the rod and reel and handline fishery off the northeast United States. ICCAT Working Doc. SCRS/93/68. 12pp.

Cushing, D.H. 1982. Climate and Fisheries. Academic Press, New York. pp.74-78, 145-155.

Cushing, D.H., and R.R. Dickson. 1976. The biological response in the sea to climate changes. Adv. Mar. Biol. 14:1-122.

De Vries, T.J., and W.G. Pearcy. 1982. Fish debris in sediments of the upwelling zone off central Peru; a late Quaternary record. Deep-Sea Research 28:87-109.

Dicenta, A., C. Piccinetti, et al. 1980. Comparison between the estimated reproductive stocks of bluefin tuna (*T. thynnus*) of the Gulf of Mexico and western Mediterranean. ICCAT Coll. Vol. Sci. Pap. IX(2):442-448 (SCRS/79/45).

Dizon, A.E., C. Cockyer, W.F. Perrin, D.P. Demaster and J. Sisson. 1992. Rethinking the stock concept: A phylogenetic approach. Conserv. Biol. 6: 24-36.

Edmunds, P.H., and J.I. Sammons. 1971. Genic polymorphism of tetrazoluim oxidase in bluefin tuna, *Thunnus thynnus*, from the western North Atlantic. J. Fish. Res. Bd. Can. 28:1053-1055.

Edmunds, P.H., and J.I. Sammons. 1973. Similarity of genetic polymorphism of tetrazolium oxidase in bluefin tuna, *Thunnus thynnus*, from the Atlantic coast of France and the western North Atlantic. J. Fish. Res. Bd. Can. 30:1031-1032.

Finnerty, J.R., and B.A. Block. 1992. Direct sequencing of mitochondrial DNA detects highly divergent haplotypes in blue marlin (*Makaira nigricans*). Molecular Marine Biology Biotechnol. 1:206-214.

REFERENCES

Finnerty, J.R., and B.A. Block. 1994. Evolution of cytochrome b in the Scombroidei (Teleostei): Molecular insights into billfish (Istiophoridae and Xiphiidae) relationships. Fishery Bull. U.S.A. In press.

Fletcher, R.I. 1975. A general solution for the complete Richards function. Math. Biosci. 27:348-360.

Food and Agriculture Organization of the United Nations. 1968. Report of the Meeting of a Group of Experts on Tuna Stock Assessment (under the FAO Expert Panel for the Facilitation of Tuna Research). FAO, Rome, p. 21.

Fujino, K., and T. Kang. 1968b. Transferrin groups of tunas. Genetics 59:79-91.

Gilbert, R.O. 1987. Statistical methods for environmental pollution monitoring. Van Nostrand-Reinhold Co., New York, 320pp.

Gran, H.H. 1931. On the conditions for the production of phytoplankton in the sea. Rapp. P.-v. Cons. int. Explor. Mer 75:37.

Grant, W.S. 1984. Biochemical population genetics of Atlantic herring, *Clupea harengus*. Copeia 1984:355-362.

Grant, W.S., and G. Stahl. 1988. Evolution of Atlantic and Pacific cod: Loss of genetic variation and gene expression in Pacific cod. Evolution 42:138-146.

Guilderson, T.P., R.G. Fairbanks, and J.L. Rubenstone. 1994. Tropical temperature variations since 20,000 years ago: modulating interhemispheric climate change. Science 263:663-665.

Hopkins, T.S., and N. Garfield III. 1979. Gulf of Maine intermediate water. J. Mar. Res. 37:103-139.

Ihssen, P.E., H.E. Booke, J.M. Casselman, J.M. McGlade, N.R. Payne, and F.M. Utter. 1981. Stock identification: Materials and methods. Can. J. Fish. Aquat. Sci. 38:1838-1855.

ICCAT. 1991. Report of the ICCAT Standing Committee on Research and Statistics (SCRS). Madrid, October 28-November 8.

ICCAT. 1992. Report of the ICCAT Standing Committee on Research and Statistics (SCRS). Madrid, November 2-6.

ICCAT. 1993. Report of the ICCAT Standing Committee on Research and Statistics (SCRS). Madrid, October 25-November 5. COM-SCRS/93/13.

Kabata, Z. and J-S Ho. 1981. The origin and dispersal of hake (genus *Merluccius:* Pisces: Teleosti) as indicated by its copepod parasites. Oceanogr. Mar. Biol. Ann. Rev. 19:381-404.

Kimura, M. 1955. Solution of a process of random genetic drift with a continuous model. Proc. Natl. Acad. Sci. USA 41:144-150.

Lo, N.C.H., D. Jacobson, and J.L. Squire. 1992. Indices of relative abundance from fish spotter data based on delta-lognormal models. Can. J. Fish. Aquat. Sci. 49:2515-2526.

Luhmann, M. 1959. Die deutsche Tunifischerei und ihre Fange in den Hahren 1956 - 1958. Arch. Fisch. Wiss. 10(1/2):68-69.

Mann, K.H., and J.R.N. Lazier. 1991. Dynamics of Marine Ecosystems, Chapter 9. Blackwell Scientific Publ., Oxford, England. pp. 342-383.

Martin, J.H. 1992. Iron as a limiting factor in oceanic productivity. In: Primary Productivity and Biogeochemical Cycles in the Sea. (eds. Falkowski and Woodhead). Plenum Press. pp. 123-137.

Mather, F.J., Jr. 1975. Trends in bluefin tuna catches in the Atlantic Ocean and the Mediterranean Sea. unpublished manuscript.

Mather, F.J. 1980. A preliminary note on migratory tendencies and distributional patterns of Atlantic bluefin tuna on recently acquired and cumulative tagging results. ICCAT Coll. Vol. Sci. Pap. IX:478-490 (SCRS/79/76).

Mather, F.J., and A.C. Jones. (1974). A preliminary review of the stock structure of bluefin tuna in the Atlantic Ocean. p.47.

Mather, F.J., J.M. Mason, and A.C. Jones. 1974. Distribution, fisheries and life history data relevant to identification of Atlantic bluefin tuna stocks. ICCAT Coll. Vol. Sci. Pap. II:234-258 (SCRS/73/54).

Murphy, G.I. 1990. A review of the evidence of stock structure in Atlantic bluefin tuna with an alternate stock hypothesis. Draft.
Nei, M. 1987. Molecular Evolutionary Genetics. Columbia University Press, New York. 512pp.
Neuparth, A.E. 1925. Note sure les fluctuations de la peche du Thon (*Thunnus thynnus*) sur la cote Sud du Portugal. Rapp. P.-v. Reun. Cons. int. Explor. Mer. 35:51-56.
National Research Council. 1994. Improving the Mangement of U.S. Marine Fisheries. National Academy Press, Washington, D.C. 62pp.
Parrack, M.L. 1990. Tagging applications-Western Atlantic. Pp. 104-109 in: D. Clay (ed.), World Bluefin Meeting, May 25-31, La Jolla, Calif.
Parrish, R.A., A. Bakun., D.M. Husby and C.S. Nelson. 1983. Comparative climatology of selected environmental processes in relation to eastern boundary current pelagic fish reproduction. In: Proceedings of the Expert Consultation to Examine Changes in Abundance and Species composition of Neritic Fish Resources. San Jose, Costa Rica, April 1983. (eds. G.D. Sharp and J. Csirke). F.A.0. Fish Rep. 291:731-777.
Phipps, M. 1980. Preliminary studies of population structure and mortality of bluefin tuna (*Thunnus thynnus*) in Nova Scotia, Canada. MSc thesis, University of Guelph.
Porch, C.E., and G.P. Scott. 1993. A numerical evaluation of GLM methods for estimating indicies of abundance from West Atlantic tuna catch per trip data when a high proportion of the trips are unsuccessful. ICCAT Working Document SCRS/93/75.
Porch, C.E., S.C. Turner, and R.D. Methot. 1993. Estimates of the abundance and mortality of West Atlantic bluefin tuna using the stock synthesis model. ICCAT Working Document SCRS/93/74.
Quinn, T.J., II, S.H. Hoag, and G.M. Southward. 1982. Comparison of two methods of combining catch-per-unit-effort data from geographic regions. Can. J. Fish. Aquat. Sci. 39:837-846.
Restrepo, V.R., C.E. Porch, S.C. Turner, G.P. Scott, and A.A. Rosenberg. 1993. Combination of spawner-recruit, spawning biomass-per-recruit and yield-per-recruit compulations for the estimation of the long term potential for West Atlantic bluefin tuna. ICCAT Working Document SCRS/93/72.
Richards, F.J. 1959. A flexible growth function for empirical use. J. Exp. Bot. 10:290-300.
Richards, W.J. 1976. Spawning of bluefin tuna (*Thunnus thynnus*) in the Atlantic Ocean and adjacent seas. ICCAT Coll. Vol. Sci. Pap. V(2):267-278 (SCRS/75/97).
Richards, W.J. 1987. Mexus-Gulf ichthyoplankton research, 1977-1984. Mar. Fish. Rev. 49(1):39-41.
Richards, W.J. 1990. Results of a review of the U.S. bluefin tuna larval assessment with a brief response. ICCAT Coll. Vol. Sci. Pap. XXXII(2):240-247 (SCRS/89/79).
Richardson P.L. 1983. Gulf Stram Rings. In: Eddies in Marine Science (ed. A.R. Robinson). Springer-Verlag, NY. pp. 19-45.
Rivas, L.R. 1954. A preliminary report on the spawning of the western north Atlantic bluefin tuna (*Thunnus thynnus*) in the Straits of Florida. Bull. Mar. Sci. Gulf Caribb. 4:302-322.
Rivas, L.R. 1978. Preliminary models of annual life history cycles of the north Atlantic bluefin tuna. In: The Physiological Ecology of Tunas. (eds. G.D. Sharp and A.D. Dizon). Academic Press, New York. pp. 369-393
Rodriquez-Roda, J. 1967. El atun, *Thunnus thynnus*, (L.) del sur de España, en la campaña almadrabera del año 1966. Invest. Pesq. 31(2):349-359.
Rodriquez-Roda, J. 1971. Investigations of tuna (*Thunnus thynnus*) in Spain. In: ICCAT Report for Biennial Period, 1970-71, Part II. pp.110-113.
Roule, L. 1924. Étude sur les déplacements et la pêche du thon (*Orcynus thynnus* L.) en Tunisie et dans la Méditerranée occidentale. Bulletin, Station Océanographique de Salammbô, Tunisia, 2:1-39.
Schnute, J. 1981. A versatile growth model with statistically stable parameters. Can. J. Fish. Aquat. Sci. 38:1128-1140.

REFERENCES

Sella, M. 1926. Altri fatti sopra le migrazioni dei tonni accertati per mezzo degli ami. Rend. Accad. Lincei (6a), 4, 2 sem., fasc. 5-6.

Sella, M. 1927. Les migrations des thons etudiees par le moyen des hameçons, Bull. Stn. Agric. Peache Castigone, fasc. 2:101-136.

Sella, M. 1929. Migrations and habitat of the tuna (*Thunnus thynnus* L.), studied by the method of the hooks, with observations on growth, on the operation of fisheries, etc. Spec. Sci. Rep. U. S. Fish Wildl. Serv. 76:20pp. (1952). Transl. by W.G. Van Campen.

Sella, M. 1931. The tuna (*Thunnus thynnus* L.) of western Atlantic. An appeal to fishermen for the collection of hooks found in tuna fish. Internationale Revue der Gesamten Hydrobiologie, Berlin, 25(1-2):46-67.

Shackleton, L.Y. 1987. A comparative study of fossil fish scales from three upwelling regions. In: The Benguela and comparable Ecosystems. (eds. A.I.L. Payne, J.A. Gulland and K.H. Brink). South African Journal of Marine Science 5:79-84.

Southward, A.J. 1980. The western English Channel—an inconstant ecosystem. Nature 285:361-366.

Squire, J.L. 1962. Thermal relationships of tuna in the oceanic northwest Atlantic. Paper presented to the World Scientific Meeting on the Biology of Tunas and Related Species, La Jolla, Calif., July 2-14. Also in FAO Fish Rep. 1963. 3(6):1639-1657.

Suzuki, Z. 1990. Reproduction-Western Atlantic. In: World Bluefin Meeting, May 25-31. (ed. D. Clay). La Jolla, Calif. pp. 97-99.

Tiews, K. 1962. Synopsis of biological data on bluefin tuna. Paper presented to the World Scientific Meeting on the Biology of Tunas and Related Species, La Jolla, Calif., July 2-14.

Tiews, K. 1963. Synopsis of biological data on bluefin tuna *Thunnus thynnus* (Linnaeus) 1758 (Atlantic and Mediterranean). Proc. World. Sci. Meeting on the Biol. of Tunas and Rel. Species, FAO Fish. Repts., 6(2):422-481.

Tiews, K. 1978. On the disappearance of bluefin tuna in the North Sea and its ecological implications for herring and mackerel. Rapp. P.-v. Reun. Cons. int. Explor. Mer. 172:301-309.

Utter, F. 1981. Biological criteria for definition of species and distinct intraspecific populations of anadromous salmonids under the U.S. Endangered Species Act of 1973. Can. J. Fish. Aquat. Sci. 38:1626-1635.

Walters, V. 1980. Ectoparasites of eastern and western Atlantic bluefin tuna. ICCAT Coll. Vol. Sci. Pap. IX(2):491-498 (SCRS/79/79).

Waples, R. S. 1991. Pacific salmon, *Oncorhynchus* spp., and the definition of "species" under the endangered species act. Mar. Fish. Rev. 53:11-22.

Watson, C., R.E. Bourke, and R.W. Brill. 1988. A comprehensive theory on the etiology of burnt tuna. Fish. Bull. 86:367-372.

Wilson, P.C., and M.R. Bartlett. 1967. Inventory of U.S. exploratory longline fishing effort and catch rates for tunas and swordfish in the northwestern Atlantic, 1957-1965. U.S. Fish and Wildlife Service Special Scientific Report—Fisheries No. 543. pp. 1-52.

Wilson, A.C., R.C. Cann, S.M. Carr, M. George, Jr., U.B. Gyllensten, K.M. Helm-Bychowski, R.G. Higuchi, S.R. Palumbi, E.M. Prager, R.D. Sage, and M. Stoneking. 1985. Mitochondrial DNA and two perspectives on evolutionary genetics. Biol. J. Linn. Soc. 26:375-400.

Wise, John P. 1991. Federal Conservation and Managment of Marine Fisheries in the United States. Washington, D.C.: Center for Marine Conservation, p 189-190.

Wyrtki, K. 1977. Sea level during the 1972 El Niño. J. Phys. Oceanogr. 7:779-87.

Valdes, A.M., M. Slatkin, and N.B. Freimer. 1993. Allele frequencies of microsatellite loci: The stepwise mutation model revisited. Genetics 133:737-749.

Yamanaka, I. and H. Yamanaka. 1970. On the variation of the current pattern in the equatorial western Pacific Ocean and its relationship to the yellowfin tuna stock. (ed. K. Sugawara). The Kuroshio II. Proc. 2nd CSK Symposium, Tokyo, Sept 28-Oct.1. pp.527-533.

Zar, J.H. 1974. Biostatistical Analysis. Prentice-Hall, Englewood Cliffs, N.J. 620pp.

Appendixes

APPENDIX
A
Biographical Sketches of Committee Members

John J. Magnuson, *Chairman*, serves as professor of zoology and director of the Center for Limnology at the University of Wisconsin, Madison. He earned his Ph.D. from the University of British Columbia in zoology. His research interests are in fish and fisheries ecology, the behavioral and distributional ecology of fishes and macroinvertebrates in lakes and oceans, community ecology of lakes as islands, ecology of the Great Lakes, and long-term ecological research on lake ecosystems, including climate change effects.

Barbara A. Block is assistant professor at the Hopkins Marine Laboratory at Stanford University. She earned her Ph.D. in comparative physiology from Duke University. Her major area of study is tuna and billfish physiology, genetics, and evolution.

Richard B. Deriso is the chief scientist of the Tuna-Billfish Program of the Inter-American Tropical Tuna Commission. He also serves as associate adjunct professor at the Scripps Institution of Oceanography and affiliate associate professor of fisheries at the University of Washington. He earned his Ph.D. in biomathematics from the University of Washington. His major research interests are in fisheries population dynamics, quantitative ecology, stock assessment, applied mathematics, and statistics.

John R. Gold is professor of genetics and director of the Center for Biosystematics and Biodiversity at Texas A&M University. He earned a Ph.D. in genetics from the University of California, Davis. His research interests are concentrated in the areas of genetics, evolution, and phylogeny of native North American fishes.

William Stewart Grant is associate professor in the Department of Genetics at Wits University in the Republic of South Africa. He earned a Ph.D. from the University of Washington, Seattle, in fishery genetics. His present research includes studies in systematics and biogeography of marine fishes.

Terrance J. Quinn II is an associate professor at the Juneau Center, School of Fisheries and Ocean Sciences at the University of Alaska, Fairbanks. He earned his Ph.D. from the University of Washington in biomathematics. His research is in the areas of fish population dynamics and biometrics.

Saul B. Saila is professor emeritus at the Graduate School of Oceanography at the University of Rhode Island. He earned his Ph.D. from Cornell University in fishery biology. His research interests are in the area of fishery biology and population dynamics. Dr. Saila has made outstanding contributions in modeling and fish population dynamics.

Lynda Shapiro is professor of biology and director of the Institute of Marine Biology at the University of Oregon. She earned her Ph.D. from Duke University. Her area of expertise is marine phytoplankton ecology. Dr. Shapiro currently serves on the NRC's Ocean Studies Board.

E. Don Stevens is professor of zoology at the University of Guelph in Canada. He earned his Ph.D. in zoology from the University of British Columbia. His expertise is in the area of the physiology, primarily of fish; mechanisms of respiration, especially as affected by muscular exercise; and comparative physiology of muscle contraction.

APPENDIX
B

Bibliography

Anderson, A.W., W.H. Stolting, and associates. 1952. Survey of the domestic tuna industry. U.S. Fish. Wildl. Serv., Special Sci. Rep. Fish. No. 104.

Bartlett, S.E., and W.S. Davidson. 1991. Identificaton of *Thunnus* tuna species by the polymerase chain reaction and direct sequence analysis of their mitochondrial cytochrome *b* genes. Can. J. Fish. Aquat. Sci. 48:309-317.

Beardsley, G.L. 1975. A review of the status of the stocks of Atlantic bluefin tuna. ICCAT Coll. Vol. Sci. Pap. III:161-172 (SCRS/74/48).

Bigelow, H.B., and W.C. Schroeder. 1953. Fishes in the Gulf of Maine. U.S. Fish Wildl. Serv., Fish. Bull. 53:338-347.

Bossert, W.[1] 1993. Review of the ICCAT bluefin tuna working group report, draft of 10/2/93, and supporting documents.

Broadhead, G.C., D.G. Chapman, S.B. Saila (Chairman), and F. Williams. 1975. Report on the status of the stock of the Atlantic bluefin tuna (*Thunnus thynnus thynnus*).

Browder, J.A. 1993. GLM analysis of medium bluefin tuna relative abundance in the western North Atlantic based on rod and reel CPUE. ICCAT Working Document SCRS/93/73.

Brown, C.A., and J.A. Browder. 1993. Standardized catch rates of small bluefin tuna in the Virginia-Rhode Island (U.S.) rod and reel fishery. ICCAT Working Document SCRS/93/67.

Brunenmeister, S. 1980. A summary and discussion of technical information pertaining to the geographical discreteness of Atlantic bluefin tuna resources. ICCAT Coll. Vol. Sci. Pap. IX(2):506-527 (SCRS/79/95).

Butterworth, D.A., and A.E. Punt. 1993. The robustness of estimates of stock status for the western North Atlantic bluefin tuna population to violations of the assumption underlying the associated assessment models. ICCAT Working Document SCRS/93/68.

Calaprice, J.R. 1986. Chemical variability and stock variation in northern Atlantic bluefin tuna. ICCAT Coll. Vol. Sci. Pap. XXIV(2):222-254 (SCRS/85/36).

---

[1] William Bossert, Harvard University, Cambridge, MA 02138.

Clay, D., ed. 1990. Atlantic bluefin tuna (*Thunnus thynnus thynnus* (L.)): A review. World Bluefin Meeting, May 25-31, La Jolla, Calif.
Clay, D. 1986. Catch and effort in the Canadian inshore bluefin tuna fishery. ICCAT Coll. Vol. Sci. Pap. XXIV:137-142 (SCRS/85/23).
Clay, D., and T. Hurlbut. 1990. Bluefin tuna (*Thunnus thynnus* L.) fisheries in Atlantic Canada: An historical review and hypothesis of minimum assemblage size. ICCAT Coll. Vol. Sci. Pap. XXXII(2):270-279 (SCRS/89/89).
Clay, D., and J.M. Porter. 1991. National Report of Canada. ICCAT, Report for the Biennial Period 1990-91, Part II, pp. 254-258.
Cort, J.L., and J. C. Rey. 1985. Analisis de los datos de marcado de atun rojo (*Thunnus thynnus* L.) en el Atlántico Este y Mediterráneo. Migracion, crecimiento, y mortalidad. ICCAT Coll. Vol. Sci. Pap. XXII:213-219 (SCRS/84/44).
Cort, J.L., and J.M. de la Serna. 1993. Revisión de los datos de marcado/recaptura de atún rojo (*Thunnus thynnus*, L.) en el Atlántico Este y el Mediterráneo. ICCAT Working Document SCRS/93/81.
Cramer, J., and G.P. Scott. 1993. Indices of abundance for large bluefin tuna, *Thunnus thynnus*, from the U.S. mandatory pelagic longline fishery in the Gulf of Mexico and off the Florida east coast. ICCAT Working Document SCRS/93/64.
Cramer, J., and S.C. Turner. 1993. Large bluefin tuna, *Thunnus thynnus*, indices of abundance from the rod and reel and handline fishery off the northeast United States. ICCAT Working Document SCRS/93/63.
DeMoraes, M.N. 1962. Development of the tuna fishery of Brazil and preliminary analysis of the first three years' data. Proc. Symp. Scombroid fishes Mondopom Comp, 1962. Symp. Ser. Biol. Assoc. India, Part III, pp. 1042-1083.
Dicenta, A., and C. Piccinetti, et al. 1980. Comparison between the estimated reproductive stocks of bluefin tuna (*T. thynnus*) of the Gulf of Mexico and Western Mediterranean. ICCAT Coll. Vol. Sci. Pap. IX(2):442-448. (SCRS/79/45) (Rev).
FAO. 1968. Report of the Meeting of a Group of Experts on Tuna Stock Assessment. FAO Fisheries Reports No. 61, 45p.
Hester, F. 1993. Factors reflecting catch and effort in the U.S. permitted fishery for Atlantic bluefin tuna. ICCAT Working Document SCRS/93/76.
Hester, F. 1993. A reexamination of the stock structure hypotheses for the Atlantic bluefin tuna. ICCAT Working Document SCRS/93/77.
ICCAT. 1990. Report of the Standing Committee on Research and Statistics (SCRS), ICCAT Report for Biennial Period, 1988-89, Part II, pp. 153-161 and 328-339.
ICCAT. Consultation on the technical aspects of methodologies which account for individual growth variability by age (St. Andrews, N.B., Canada, July 6-10, 1993). SCRS/93/17.
Keyranfar, A. 1962. Serologie et immunolgre de deux especies de thonides (Germa alalunga Ginelin et Thunnus thynnus Linne) de L'Atlantique et de la Mediterranee. Rev. Trav. Inst. Peches Marit. 26(4):407-456.
Mather, F.J. 1975. Trends in bluefin tuna catches in the Atlantic Ocean and the Mediterranean Sea. Unpublished manuscript.
Mather, F.J., J.M. Mason, et al. 1974. Distribution, fisheries and life history data relevant to identification of Atlantic bluefin tuna stocks. ICCAT Coll. Vol. Sci. Pap. II:234-258 (SCRS/73/54).
Miyabe, N. 1993. Trends of CPUE for Atlantic bluefin tuna caught by the Japanese longline fishery up to 1992. ICCAT Working Document SCRS/93/48.
Miyabe, N., and K. Hirumatsu. 1993. Description of the Japanese longline fishery operating in the central North Atlantic. ICCAT Working Document SCRS/93/49.
Miyake, P.M. 1971. Statistical Bulletin. Inter. Comm. Cons. Atlantic Tunas, Madrid. (ST/CA/71/3-2).

Miyake, P.M., and P. Kebe. 1993. Procedures adopted in preparing data on west Atlantic bluefin tuna catch at size for 1993 SCRS meeting. ICCAT Working Document SCRS/93/7.

Murphy, G.I. 1990. A review of the evidence of stock structure in Atlantic bluefin tuna with an alternate stock hypothesis. Draft.

Neuparth, A.E. 1925. Note Sur Les Fluctuations de la Pêche du Thon (*Thunnus thynnus* L.) Sur La Côte Sud du Portugal. Rapp. Proes-Verb. Reun. 35:51-56.

Parrack, M.L. 1981. An Assessment of the Atlantic Bluefin Tuna. ICCAT Coll. Vol. Sci. Pap. XV(2):209-221 (SCRS/80/43).

Parrack, M.L. 1982. Atlantic Bluefin Tuna Resource Update. ICCAT Coll. Vol. Sci. Pap. XVII(2):315-328 (SCRS/81/55).

Parrack, M.L. 1986. A Method of Analyzing Catches and Abundance Indices from a Fishery. ICCAT Coll. Vol. Sci. Pap. XXIV:209-221 (SCRS/85/35).

Porch, C.E., and G.P. Scott. 1993. A numerical evaluation of GLM methods for estimating indices of abundance from West Atlantic bluefin tuna catch per trip data when a high proportion of the trips are unsuccessful. ICCAT Working Document SCRS/93/75.

Porch, C.E., S.C. Turner, and R.D. Methot. 1993. Estimates of the abundance and mortality of West Atlantic bluefin tuna using the stock synthesis model. ICCAT Working Document SCRS/93/74.

Porter, J.M. 1992. National Report of Canada. ICCAT, Report for the Biennial Period 1992-93, Part 1, pp. 341-346.

Porter, J.M. 1993. National Report of Canada. ICCAT Working Document SCRS/93/53, in press.

Porter, J.M., and H. Stone. 1993. A mark-recapture experiment on bluefin tuna from the Browns-Georges Banks region of the Canadian Atlantic: 1993 update. ICCAT Working Document SCRS/93/50.

Prager, M.H., and G.P. Scott. 1993. A nonequilibrium production model of bluefin tuna in the western North Atlantic Ocean. ICCAT Working Document SCRS/93/71.

Restrepo, V.R., C.E. Porch, S.C. Turner, G.P. Scott, and A.A. Rosenberg. 1993. Combination of spawner-recruit, spawning biomass-per-recruit and yield-per-recruit compultations for the estimation of the long term potential for west Atlantic bluefin tuna. ICCAT Working Document SCRS/93/72.

Rodriquez-Roda, J. 1967. El atun, *Thunnus thynnus*, (L.) del sur de Espana, en la campana almadrab-ora del ano 1966. Invest. Resq. 31(2):349-359.

Rodriquez-Roda, J. 1971. Investigations of tuna (*Thunnus thynnus*) in Spain. Pp. 110-113 in: ICCAT Report for biennial period, 1970-1971, Part II.

Sakagawa, G.T. 1975. The purse-seine fishery for bluefin tuna in the northwestern Atlantic Ocean. MFR Paper 1126. Mar. Fish. Rev. 37(3):1-8.

Sakagawa, G.T., and A.L. Coan. 1974. A review of some aspects of the bluefin tuna (*Thunnus thynnus thynnus*) fisheries of the Atlantic Ocean. ICCAT Coll. Vol. Sci. Pap. II:259-313 (SCRS/73/60).

Scott, G.P., and S.C. Turner. 1993. An updated index of West Atlantic bluefin spawning biomass based on larval surveys in the Gulf of Mexico. ICCAT Working Document SCRS/93/69.

Suda, A. 1993. Examination of the relationship between Mid-Atlantic and northwest Atlantic bluefin tuna concentrations. ICCAT Working Document SCRS/93/138.

Suzuki, S. 1993. Proposal of some important researches on the Atlantic bluefin tuna. ICCAT Working Document SCRS/93/47.

Thompson, H.C., Jr., and R.F. Contin. 1980. Electrophoretic study of Atlantic bluefin tuna (*Thunnus thynnus*) from the eastern and western north Atlantic Ocean. ICCAT Coll. Vol. Sci. Pap. IX(2):499-505 (SCRS/79/96).

Tiews, K. 1963. Synopsis of biological data on bluefin tuna *Thunnus thynnus* (Linneaus) 1758 (Atlantic and Mediterranean). Proceedings of the World Scientific Meeting on the biology of tunas and related species. FAO Fish. Rep., 6(2):422-481.

Tiews, K. 1978. On the disappearance of bluefin tuna in the North Sea and its ecological implications for herring and mackerel. Rapp. P.v. Reun. Cons. Int. Explor. Mer, 172:310-309.

Turner, S.C. 1986. An analysis of recaptures of tagged bluefin with respect to the mixing assumption. ICCAT Coll. Vol. Sci. Pap. XXIV:196-202 (SCRS/85/33) (Rev).

Turner, S.C., and V.R. Restrepo. 1993. A review of the growth rate of West Atlantic bluefin tuna, *Thunnus thynnus*, estimated from marked and recaptured fish. ICCAT Working Document SCRS/93/65.

Turner, S.C., and M. Terceiro. 1993. Estimation of West Atlantic bluefin tuna, *Thunnus thynnus*, age composition with length compostition analysis. ICCAT Working Document SCRS/93/66.

Turner, S.C., C.E. Porch, and V.R. Restrepo. 1993. Sensitivity of projections of west Atlantic bluefin tuna stock size to catches in the central Atlantic region and to retrospective patterns in historical stock size estimates. ICCAT Working Document SCRS/93/121.

Westman, J.R., and P.W. Gilbert. 1941. Notes on age determination and growth of the Atlantic bluefin tuna, *Thunnus thynnus* (Linnaeus). Copeia 1941:70-73.

Westman, J.R., and W.C. Neville. 1942. The tuna fishery of Long Island, New York. U.S. Fish Wildl. and County Board of Nassau County. pp.1-31.

Wilson, P.C. 1965. Review of the development of the Atlantic Coast tuna fishery. Comm. Fish. Rev. 27(3):1-10.

Wilson, P.C., and M.R. Bartlett. 1967. Inventory of U.S. exploratory longline fishing effort and catch rates for tunas and swordfish in the northwestern Atlantic, 1957-65. U.S. Fish Wildl. Serv., Spec. Sci. Rep. Fisheries No. 543.

Wise, J.P., and C.W. Davis. 1973. Seasonal distribution of tunas and billfishes in the Atlantic. NOAA Technical Report NMFS SSRF-662, 24pp.

Woodley, C.M. 1993. Determination of stock structure in bluefin tuna at the NMFS Laboratory, Charleston, SC. ICCAT Working Document SCRS/93/62.

APPENDIX
C
Genetic Variation in Other Tunas and Related Fish[1]

Tuna populations have been studied with immunologically detectable blood markers, protein electrophoresis, and restriction enzyme analysis and direct sequencing of mtDNA. The first genetic study of tuna (Cushing, 1956) demonstrated variability in red blood cell antigens that are analogous to human blood types. Subsequently, several workers (Suzuki, 1962; Fujino, 1970; Suzuki et al., 1958, 1959) attempted to assay blood group variability with immunological methods to study population structure but were largely unsuccessful in resolving intraocean population structure. More recently, protein electrophoresis and mtDNA restriction enzyme analysis have been used to search for population differences. The results for three tunas, yellowfin (*Thunnus albacares*), albacore (*Thunnus alalunga*), and skipjack (*Katsuwonus pelamis*), are reviewed here because of the similarity of these species to bluefin tuna, and because the results may give insight into the possible population genetic structure of bluefin tuna.

YELLOWFIN TUNA

Scoles and Graves (1993) examined RFLPs of mtDNA from 20 yellowfin tuna sampled from each of five widely spread Pacific Ocean localities ($n = 100$) and one Atlantic Ocean locality ($n = 20$). Although they found high levels of

---

[1]All statistical results discussed in this appendix were recalculated with the log-likelihood ratio statistic (G-test) from data given in the original articles, because the G-test is now preferred over the chi-square test used in the original studies.

genetic diversity, there was no evidence of genetic differentiation among samples; common haplotypes occurred in similar frequencies in all samples. These results are consistent with high levels of gene flow among localities throughout the Pacific Ocean and between Pacific and Atlantic Ocean localities. Ward et al. (1994) examined four polymorphic loci encoding allozymes and mtDNA with two informative restriction enzymes among seven samples from the western, central, and eastern Pacific Ocean, and a fifth polymorphic enzyme in eastern and central Pacific samples. They found no significant frequency differences among localities in mtDNA haplotypes or allozymes at four loci. The frequencies for GPI-F, however, were not significantly different between two eastern Pacific samples (southern California and southern Mexico), but were significantly different between these samples and samples from the central and western Pacific Ocean (Coral Sea, Philippines, Kiribati) and two samples taken near Hawaii. Their results were consistent with an earlier study (Sharp, 1978) in identifying heterogeneity between these areas. Ward et al. (1994) concluded that gene flow between eastern and western Pacific yellowfin tunas was severely restricted, with only a few individuals per generation moving between the two regions. The lack of concordance of the other four polymorphic loci and the mtDNA haplotypes with the PGI-F locus, suggests minimally that additional study of mtDNA in yellowfin tuna is warranted. The study of Ward et al. (1994) does emphasize the need for multiple molecular genetic techniques for examining the stock structure of a given species (i.e., the same conclusion may not have been reached with the use of any one technique).

ALBACORE TUNA

The combined results of Suzuki et al. (1958, 1959), Suzuki (1962), and Fujino (1970) for the Tg blood group of albacore tuna showed little allele-frequency heterogeneity between albacore tuna sampled from the north and south Pacific Ocean, suggesting that fish in this area consist of a single, genetically homogeneous population. There were, however, allele-frequency shifts between samples from the Atlantic and Indian Oceans, indicating population-level differentiation. In another study of albacore tuna, Keyvanfar (1962) found significant frequency differences in blood group alleles between samples from the Atlantic Ocean and the Mediterranean Sea. He also found qualitative immunodiffusion differences between albacore tuna from the Atlantic Ocean and Mediterranean Sea; Atlantic fish had an antigen that was apparently lacking in Mediterranean fish. The genetic basis of this difference is unknown. Graves and Dizon (1989) analyzed mtDNA between albacore tuna from southern Africa ($n = 11$) and San Diego ($n = 12$). They found six fragment length variants in individual fish but virtually no differentiation between samples from the Atlantic and Pacific Oceans. The high proportion of shared haplotypes is strong evidence for recent or ongoing gene flow between oceans. Similar results have recently been obtained with other highly migratory fishes that inhabit temperate seas (Graves and McDowell, 1994). However, given the small sample size of the study, the results are not definitive.

SKIPJACK TUNA

Skipjack tunas (*Katsuwonus pelamis*) are more tropical and have a more restricted temperature range than bluefin tuna. An analysis of the allelic frequencies of transferrin in skipjack tuna (Fujino and Kang 1968b) revealed a significant difference ($G = 22.12$, degrees of freedom = 2, $P < 0.01$) between samples from Hawaii ($n = 2,257$) and the eastern Pacific Ocean ($n = 175$). There were no significant differences, however, between samples from the eastern ($n = 2,432$) and western ($n = 3,792$) Pacific Ocean or between samples from the Atlantic ($n = 213$) and Pacific ($n = 4,328$) Oceans. Richardson (1983) made an extensive study of 42 isozyme loci in 70 skipjack samples collected throughout the Pacific Ocean and found a longitudinal cline across the central and southwestern Pacific Ocean for an esterase locus and a locus encoding guanine deaminase. The distribution of allelic frequencies appeared to fit an isolation-by-distance model of migration, and estimates of genetic neighborhood size were about 2,000km. This estimate was similar to neighborhood sizes estimated from tagging data. These results indicate that the dispersal range of skipjack is restricted enough to allow genetic differences to accumulate among regions in the Pacific Ocean.

Two studies have searched for differences between Pacific and Atlantic skipjack populations. Fujino (1969) concluded from the analysis of serum esterase frequencies and three blood group systems that skipjack tuna from the Pacific Ocean ($n = 1,080$) were distinct genetically from those in the Atlantic Ocean ($n = 127$). More recently, Graves et al. (1984) used nine restriction enzymes to search for mtDNA sequence variability in skipjack tuna from Hawaii ($n = 9$), Brazil ($n = 6$), and Puerto Rico ($n = 1$). Although they found polymorphisms within each ocean, sequence divergence between samples from the Atlantic and Pacific Oceans was 0.0% (i.e., there were no demonstrable mtDNA differences). However, the number of individuals examined by Graves et al. (1984) was limited.

BILLFISHES

Recent studies on the population genetics of billfishes (Xiphiidae and Istiophoridae) provide a basis for comparison, given that many species of billfishes are highly migratory and occupy pelagic habitats similar to those occupied by tunas. Finnerty and Block (1992) sequenced 612 bp of the cytochrome *b* gene from 26 blue marlin (*Makaira nigricans*) from the Pacific Ocean ($n = 14$) and the Atlantic Ocean ($n = 12$) and found two distinct lineages (Pacific Ocean or ubiquitous and Atlantic Ocean only) that differed by at least nine substitutions. Maximum-parsimony analysis of blue marlin cytochrome *b* variants revealed the occurrence of two major evolutionary lines. The frequencies of the haplotypes in the Atlantic Ocean and Pacific Ocean samples were significantly different ($P < 0.05$). Direct sequencing of the control region of mtDNA, a region of higher variability, provided a similar result: a ubiquitous blue marlin mtDNA lineage found in both ocean basins and a unique mtDNA lineage specific to the Atlantic Ocean ($n = 40$ [Meyer et al., unpublished results]). Graves and McDowell (1994) examined mtDNA sequence variability using RFLPs among Atlantic ($n = 56$) and Pacific ($n = 58$) samples. A single group of closely related haplotypes was found among samples from different oceans, and a discrete mtDNA haplotype that differed by several restriction sites changes was identified in a subset of the Atlantic samples. A similar result was found in another species of warm temperate to tropical billfish, the sailfish (*Istiophorus platypterus*). In blue marlin, two distinct evolutionary lineages with historical roots in the separate ocean basins (Pacific Ocean versus Atlantic Ocean) are hypothesized to be associated with formation of the land bridge between ocean basins and the temperature distribution of the species whch limits intermixing to South African waters (Finnerty and Block, 1992).

The striped marlin (*Tetrapturus audax*), a cold temperate marlin thought to be restricted to the Pacific Ocean basin, has been examined in three studies (Block et al., 1993; Finnerty and Block, 1994, Graves and McDowell,[2] unpublished data). A surprising result has been the inability to genetically distinguish the striped marlin from the morphologically distinct white marlin (*Tetrapturus albidus*), a temperate-water marlin of the Atlantic Ocean. Direct sequencing of 612 bp of the cytochrome *b* gene in striped marlin ($n = 2$) from the Pacific Ocean and white marlin ($n = 2$) from the Atlantic Ocean revealed a sequence divergence of <0.1%. In a larger study, Graves and McDowell (1994) used mtDNA RFLP analysis of white and striped marlin and found no mtDNA differences between the two species. Within the Pacific Ocean basin, Graves and McDowell (1994) examined samples taken from four localities in the Pacific Ocean (near Mexico, Ecuador, Australia, and Hawaii) and found significant differences among the frequencies of composite mtDNA haplotypes. The significant heterogeneity evident in the striped marlin sample may be due to differences in the absolute number of migrants exchanged between ocean populations or an underlying behavioral difference associated with fidelity to a spawning ground in this species.

Swordfish (*Xiphias gladius*) are ecologically distinct from istiophorid billfishes and have a cosmopolitan distribution and wider temperature tolerance due to a unique endothermic strategy (Carey, 1982; Block et al., 1993). Tag and

---

[2]John E. Graves and Jan R. McDowell, Virginia Institute of Marine Science, School of Marine Science, College of William and Mary, Glouchester Point, VA 23062.

release programs have not revealed extensive large-scale transoceanic migratory movements, but instead have demonstrated predominantly latitudinal north and south movements. Recent studies have attempted to examine the genetic structure of swordfish. Magoulas et al. (1992) used two informative restriction enzymes to examine mtDNA variation among three samples of swordfish from the Mediterranean Sea (near Spain, $n = 94$; near Greece, $n = 73$; and near Italy, $n = 75$), two samples from the eastern Atlantic Ocean (near Gibralter, $n = 40$), and one sample from the Gulf of Guinea ($n = 95$). The Mediterranean samples did not differ significantly from one another in mtDNA genotype frequency. The sample from the Gulf of Guinea differed significantly in haplotype frequencies from all other samples, minimally suggesting the existence of two discrete swordfish populations. Grijalva-Chon et al. (in press) surveyed mtDNA RFLP variability in samples from the western ($n = 42$), central ($n = 42$), and eastern ($n = 59$) Pacific Ocean and found no significant difference among haplotypic frequencies. Rosel[3] and Block (unpublished results) sequenced the D-loop region from a worldwide sample of 150 swordfish and found evidence for both global and interocean mixing of populations as well as discrete clades unique to ocean basins.

Molecular genetic studies of istiophorid billfishes (marlin and sailfish) have revealed more genetic structure than for tunas. In contrast, swordfish throughout the Atlantic and Pacific Oceans have a more homogeneous genetic stock structure. This observed difference in population structuring among istiophorid billfishes, swordfish, and tunas may reflect differences in exchange rates between ocean basins that ultimately may be associated with the thermal ecology/physiology of each species (many of the tuna species and swordfish have endothermic capabilities that allow them to move through wider temperature gradients). In the case of the swordfish, *Xiphias gladius*, controversy exists as to whether there are one or two stocks in the Atlantic Ocean. In contrast to bluefin tuna, however, several molecular genetic studies have been conducted in the past three years (see preceding paragraph).

REFERENCES

Alvarado-Bremer, J.R. 1992. Stock differentiation of Atlantic swordfish using mitochondrial DNA analysis. ICCAT Coll. Vol. Sci. Pap. XXXIX(2):607-614 (SCRS/91/48) (Rev).

Baglin, Jr., R.E. 1982. Reproductive biology of western Atlantic bluefin tuna. Fish. Bull. 80: 121-134.

Bartlett, S.E., and W.S. Davidson. 1991. Identification of *Thunnus* tuna species by the polymerase chain reaction and direct sequence analysis of their mitochondrial cytochrome *b* genes. Can. J. Fish. Aquat. Sci. 48: 309-317.

---

[3]Patty Rosel, Hopkins Marine Station, Stanford University, Oceanview Blvd., Pacific Grove, CA 93950.

Block, B.A., J.R. Finnerly, A.F.R. Stewart, and J. Kidd. 1993. Evolution of endothermy in fish: Mapping physiological traits on a molecular phylogeny. Science 260: 210-214.

Brunemeister, S. 1980. A summary and discussion of technical information pertaining to the geographical discreetness of Atlantic bluefin tuna resources. ICCAT Coll. Vol. Sci. Pap. IX:506-527 (SCRS/79/95).

Calaprice, J.R. 1986. Chemical variability and stock variation in northern Atlantic bluefin tuna. ICCAT Coll. Vol. Sci. Pap. XXIV(2):222-254 (SCRS/85/36).

Calaprice, J.R., H.M. McShefrey, and L.A. Lapi. 1971. Radioisotope X-ray fluorescence spectrometry in aquatic biology: a review. J. Fish. Res. Board Can. 28: 1583-1594.

Carey, F.G. 1982. A brain heater in Swordfish. Science 216, p.1327(3).

Chow, S., and S. Inoue. 1993. Intra- and interspecific restriction fragment length polymorphism in mitochondrial genes of *Thunnus* tuna species. Bull. Nat. Res. Inst. Far Seas Fish. 30: 207-225.

Clay, D., ed. 1990. Atlantic bluefin tuna (*Thunnus thynnus* (L.)): A review. World Bluefin Meeting, May 25-31, La Jolla, Calif. pp. 89-180.

CLIMAP. 1976. The surface of the ice-age earth. Science 191: 1131-1136.

Committee on Fisheries. 1994. Improving the management of U.S. marine fisheries. National Academy Press. 82 pp.

Cort, J.J., and B. Liorzou. 1990a. Larval biology - Eastern Atlantic and Mediterranean. In: World Bluefin Meeting, May 25-31. (ed. D. Clay). LaJolla, Calif. p. 95

Cort, J.J., and B. Liorzou. 1990b. Reproduction - Eastern Atlantic and Mediterranean. In: World Bluefin Meeting, May 25-31. (ed. D. Clay). La Jolla, Calif. pp. 99-101.

Cort, J.J., and B. Liorzou. 1990c. Tagging interpretation - Eastern Atlantic and Mediterranean. In: World Bluefin Meeting, May 25-31. (ed. D. Clay). La Jolla, Calif. pp. 110-127.

Cort, J.L., and J.M. de la Serna. 1993. Revision de los datos de marcado/recaptura de atun roho (*Thunnus thynnus* L.) en el Atlantico Este y Mediterraneo. ICCAT .

Cushing, J.E. 1956. Observations on serology of tuna. U.S. fish Wildl. Serv., Spec. Sci. Rept. Fish. 183. 14 pp.

Dicenta, A., C. Piccinetti, et al. 1980. Comparison between the estimated reproductive stocks of bluefin tuna (*T. thynnus*) of the Gulf of Mexico and western Mediterranean. ICCAT Coll. Vol. Sci. Pap. IX:442-448 (SCRS/79/45).

Dizon, A.E., C. Cockyer, W.F. Perrin, D.P. Demaster, and J. Sisson. 1992. Rethinking the stock concept: a phylogenetic approach. Conserv. Biol. 6: 24-36.

Edmunds, P.H. and J.I. Sammons, 111. 1971. Genic polymorphism of tetrazolium oxidase in bluefin tuna, *Thunnus thynnus*, from the western North Atlantic. J. Fish. Res. Board Can. 28: 1053-1055.

Edmunds, P.H. and J.I. Sammons, 111. 1973. Similarity of genic polymorphisms of tetrazolium oxidase in bluefin tuna, *Thunnus thynnus* from the Atlantic Coast of France and the western North Atlantic. J. Fish. Res. Board Can. 30: 1031-1032.

Finnerty, J.R. and B.A. Block. 1992. Direct sequencing of mitochondrial DNA detects highly divergent haplotypes in blue marlin (*Makaira nigricans*) Mol. Mar. Biol. Biotechnol. 1: 206-214.

Fujino, K. 1969. Atlantic skipjack tuna genetically distinct from the Pacific specimens. Copeia 1969(3): 626-629.

Fujino, K. 1970. Skipjack subpopulation identified by genetic characteristics in the western Pacific. Proc. CSK Symp., East-West center, Honolulu, Hawaii, April 29-May 2, 1968.

Fujino, K. 1970. Immunological and biochemical genetics of tunas. Trans. Am. Fish. Soc. 99(1): 152-178.

Fujino, K., and T. Kang. 1968a. Serum esterase groups of Pacific and Atlantic tunas. Copeia 1968(1): 56-63.

Fujino, K., and T. Kang. 1968b. Transferrin groups of tunas. Genetics 59: 79-91.

Grant, W.S. 1984. Biochemical population genetics of Atlantic herring, *Clupea harengus*, Copeia 1984: 355-362.

Grant, W.S. and G. Stahl. 1988. Evolution of Atlantic and Pacific cod: Loss of genetic variation and gene expression in Pacific cod. Evolution 42: 138-146.

Graves, J.E., S.D. Ferris, and A.E. Dizon. 1984. Close genetic dimilarity of Atlantic and Pacific skipjack tuna (*Katsuwanus pelamis*) demonstrated with restriction endonuclease analysis of mitochondrial DNA. Mar. Biol. 79: 315-319.

Graves, J.E., and A.E. Dizon. 1989. Mitochondrial DNA sequence similarity of Atlantic and Pacific albacore tuna (*Thunnus alaluna*) Can. J. Fish. Aquat. Sci. 46: 870-873.

Graves, J.E. and J.R. McDowell. 1994. Genetic analysis of striped marlin *Tetrapturus audax* population structure in the Pacific Ocean. Can. J. Fish. Aquat. Sci. In press.

Graves, J.E. and J.R. McDowell. submitted. Inter-Ocean genetic divergence of istiophorid billfishes.

Grijalva-Chon, J.M., K. Numachi, 0. Sosa-Nishizaki, and J. de la Rosa-Velez. 1994. Mitochondrial DNA analysis of north Pacific swordfish (*Xiphias gladius*) population structure. Mar. Ecol. Prog. Ser., in press.

Gunderson, T.P., R.G. Fairbanks, and J.L. Rubenstone. 1994. Tropical temperature variations since 20,000 years ago: modulating interhemispheric climate change. Science 263:663-665.

Gutierrez. M. 1967. Estudio hematologicos en el atun, *Thunnus thynnus* (L.). Immunologia: Grupos sanguineos, eritrocitos y fitoaglutininas. Invest. Pesq. 31:137-143.

Ihssen, P.E., H.E. Booke, J.M. Casselman, J.M. McGlade, N.R. Payne, and F.M. Utter. 1981. Stock identification: Materials and Methods. Can. J. Fish. Aquat. Sci. 38:1838-1855.

Keyvanfar, A. 1962. Serologie et immunoligie de deux especes de thonides (*Germo alalunga* Gmelin et *Thunnus thynnus* Linne) de l'atlantique et del la Mediterranee. Rev. Trav. Inst. Peches. Mar. 26(4): 407-456.

Kimura, M. 1955. Solution of a process of random genetic drift with a continous model. Proc. Natl. Acad. Sci. USA 41: 144-150.

Lee, J.Y. 1965. Observations sur la serologie et l'immunologie des thons rouges (*Thunnus thynnus* Linne) de Mediterranee. Rapp. Proc.-Ver. Reunions Comm. Int. Explor. Sci. Mer Mediter. 18: 225-228.

Magoulas, A., G. Kotoulas, J.M. de la Sema, G. de Metrio, N. Tsimendides, and E. Zouros. 1992. Genetic structure of swordfish (*Xiphias gladius*) populations of the Mediterranean and the eastern side of the Atlantic: Analysis by mitochondrial DNA markers. ICCAT Coll. Vol. Sci. Pap. XL(1):126-136 (SCRS/92/84).

Mather, F.J. 1980. A preliminary note on migratory tendencies and distributional patterns of Atlantic bluefin tuna based on recently accquired and cumulative tagging results. ICCAT Coll. Vol. Sci. Pap. IX(2):478-491 (SCRS/79/76).

Mather, F.J., J.M. Mason, and A.C. Jones. 1974. Distribution, fisheries and life history data relevant to identification of Atlantic bluefin tuna stocks. ICCAT Coll. Vol. Sci. Pap. II:234-258 (SCRS/73/54).

Murphy, G.I. 1990. A review of the evidence of stock structure in Atlantic bluefin tuna with an altemate stock hypothesis. Unpbl. MS, Div. Environ. Stud., Univ. California, Davis, California, USA.

Parrack, M.L. 1990. Tagging applications - Western Atlantic. In: World Bluefin Meeting, May 25-31. (ed. D. Clay). La Jolla, Calif. pp. 104-109.

Phipps, M. 1980. Preliminary studies of population structure and mortality of bluefin tuna (*Thunnus thynnus*) in Nova Scotia, Canada. MSc. thesis, University of Guelph.

Richards, W.J. 1976. Spawning of bluefin tuna (*Thunnus thynnus*) in the Atlantic Ocean and adjacent seas. ICCAT Coll. Vol. Sci. Pap. V(2):267-278 (SCRS/75/97).

Richards, W.J. 1987. Mexus-Gulf ichthyoplankton research, 1977-1984. Mar. Fish. Rev. 49: 39-41.

Richards, W.J. 1990. Early life history - Western Atlantic. In: World Bluefin Meeting, May 25-31, (ed. D. Clay). La Jolla, Calif. pp. 93-94.

Richardson, B.J. 1983. Distribution of protein variation in skipjack tuna (*Katsuwonus pelamis*) from the central and southwestern Pacific. Australian Journal of Marine and Freshwater Research. 34:231-251.

Rodriquez-Roda, J. 1967. Fecundidad del atun, *Thunnus thynnus* (L.), de la costa sudatlantica de Espana. Invest. Pesq. 31: 33-52.

Rodriquez-Roda, J. 1971. Investigations of tuna (*Thunnus thynnus*) in Spain. ICCAT Report for biennial period, 1970-1971, Part 11: 110-113.

SCRS. 1992. Report of the ICCAT Standing Committee on Research and Statistics, Selected pages on bluefin tuna.

SCRS. 1993. Report of the ICCAT Standing Committee on Research and Statistics, Selected pages on bluefin tuna.

Scoles, D.R., and J.E. Graves. 1993. Genetic analysis of the population structure of yellowfin tuna, *Thunnus albacares*, from the Pacific Ocean. Fish. Bull. 91: 690-698.

Sharp, G.D., and A.E. Dizon. 1978. The physiological ecology of tunas proceedings of the tuna physiology workshop. Jan. 10-15, 1977. La Jolla, CA. 485pp.

Smith, A. C. 1962. The electrophoretic characteristics of albacore, bluefin tuna, and kelpbass eye-lens proteins. Calif. Fish Game 48(3): 199-201.

Smith, A. C. 1965. Intraspecific eye-lens protein differences in yellowfin tuna, *Thunnus albacores*. Calif. Fish Game 51(3): 163-167.

Smith, A. C. 1966. Electrophoretic studies of solubleprotein from lens-nuclei of bluefin tuna, *Thunnus thynnus*, from California and Australia. Amer. Zool. 6: 333 (Abstr.)

Sprague, L. M. 1962. Blood group studies of albacore (*Germo alualunga*) tuna from the Pacific Ocean, p. 37. In: Pacific Tuna Biology Conference, Aug. 14-19, 1961. Honolulu, Hawaii (ed. J.C. Marr). U.S. Fish Wildl. Serv., Spec. Sci. Rept. Fish. 415. (Abstr.) p. 37.

Suzuki, A. 1962. Serological studies of the races of tuna VI. Bigeye-3 antigens occurred in the albacore. Rep. Nankai Reg. Fish. Res. Lab. 16: 67-70.

Suzuki, A., T. Morio, and K. Mimoto. 1959. Serological studies of the races of tuna 11. Blood group frequencies of the albacore in Tg system. Part 1. Comparison of the Indian and the northwestern Pacific Ocean. Rep. Nankai Reg. Fish. Res. Lab. 11: 17-23.

Suzuki, A., Y. Shimizu and T. Morio. 1958. Serological studies of the races of tuna 1. The fundamental investigations and the blood groups of albacore. Rep. Nankai Reg. Fish. Res. Lab. 8: 104-116.

Suzuki, Z. 1990. Reproduction - Western Atlantic. In: World Bluefin Meeting, May 25-31. (ed. D. Clay). La Jolla, Calif. pp. 97-99.

Tiews, K. 1957. Biologische Untersuchungen am Roten Thun (*Thunnus thynnus*, Linnaeus) in der Nordsee. Ber. Dtsch. Wiss. Komm. Meeresforsch. 14(3): 192-220.

Tiews, K. 1963. Synopsis of biological data on bluefin tuna *Thunnus thunnus* (Linnaeus) 1758 (Atlantic and Mediterranean). FAO Fish. Biol. Synop. No. 56, pp. 422-481.

Utter, F. 1981. Biological criteria for definition of species and distinct intraspecific populations of anadromous salmonids under the U.S. Endangered Species Act of 1973. Can. J. Fish. Aquat. Sci. 38: 1626-1635.

Walters, V. Ectoparasites of eastern and western Atlantic bluefin tunas. ICCAT Coll. Vol. Sci. Pap. IX(2):491-498 (SCRS/79/82).

Waples, R. S. 1991. Pacific salmon, *Oncorhynchus* spp., and the definition of "species" under the endangered species act. Mar. Fish. Rev. 53: 11-22.

Ward, R. D., N. G. Elliott, P. M. Grewe, and A. J. Smolenski. 1994. Allozyme and mitochondrial DNA variation in yellowfin tuna (*Thunnus albacores*) from the Pacific Ocean. Mar. Biol. 118:531-539.

APPENDIX
D

Archival Tag Technology

A novel type of tagging technology provides a new and promising avenue for obtaining data on the movements of bluefin tuna in the Atlantic Ocean. Until recently, two types of tagging technologies have provided a limited picture of the movements and behavior of individual bluefin tuna in the Atlantic Ocean. Data obtained from conventional tag and release programs are vital for discerning the rate of movement and the extent of mixing in the Atlantic Ocean. Acoustic or sonic tags have provided information on daily behavior, depth and temperature preference, and body temperature. Acoustic tags provide data at a time resolution of seconds for periods of up to five days. The acquisition of such data is technically challenging and requires extensive ship time. Recently, a new type of tag called an archival tag has been developed by separate engineering efforts in four nations. The tag offers a powerful tool for discerning the movements, geoposition, and behavior of individual highly migratory fishes. The tags will get wide use in the coming five years in fisheries that have a proven tag and recapture program (the archival tag must be retrieved to obtain data).

Archival tags are miniature data loggers that have been specifically designed for use on fishes, sea turtles, and marine mammals. The tags are currently being used on tunas, plaice, salmon, whale sharks, sea turtles, and seals to study behavior, movement, and physiology. The archival tag records and stores environmental and behavioral data. In fish, the tags are designed to be returned like conventional tags, via the commercial and sportfishery. A series of sensors provides information on temperature, depth, and irradiance. The recording of irradiance can establish the time of sunrise and sunset in each day as well as the day length that is currently being used for geopositioning. Software provides graphical representation of all data. Tags currently on the market weigh 25 g in air, have up to one megabyte of memory, can retain data for 20 years, and have a lifetime of four to five years. The tags cost approximately $1,000 each. Development teams are concentrating on reducing the size of the package, increasing the memory, and reducing the cost. An additional area of experimentation is in the attachment methodologies. The requirement for long-term attachment necessitates development of attachment methods that can hold the package in place for years. Methods for mounting the tags will vary widely, depending on the species being examined.

Archival tags typically contain a data logger board with a microcontroller, a mass data store with up to one megabyte of memory, and an analog sensor board. The tags vary between development teams but range from having one to seven sensors. This flexibility is designed to allow the user to be able to tailor the tag to the different requirements for studies on a given species. Tags are coated in an epoxy resin to keep the weight of the unit as low as possible. Programming and data retrieval are done with PC-based software. Communication is via a serial RS232 infrared optical link in both the English and Australian tags (the most advanced available).

The Australians have recently put archival tags on bluefin tuna. Use of the tag followed an extensive conventional and acoustic tagging effort that provided data suggesting that archival tagging would be successful. From 1990 to 1993, over 27,000 southern bluefin tuna were tagged with conventional techniques. To date, over 900 have been returned via Australian and Japanese fisheries. Acoustic tagging efforts have provided approximately 16 days of depth and temperature preference data from 11 individual bluefin tuna. The success of handling the fish in both programs with positive results (returns and survival upon acoustic tag and release) led to a recent experiment with archival tagging on one- and two-year-old bluefin tuna. Archival tags have been placed in numerous bluefin tuna over the past year. The tags are inserted invasively with the light sensor trailing outside the fish. To date, three tags have been returned with several months of data. Implementation of this program with successful attachment and recovery bodes well for the use of this technology on the northern bluefin tuna in the Atlantic Ocean. The conventional tag and release program in the Atlantic Ocean provides encouraging numbers which indicate that archival tagging would be successful in the Atlantic Ocean. The recovery of just a few tags would exponentially increase our knowledge of where these fish go, and a major effort in this area should be encouraged. Problems that need to be solved before archival tags are put on bluefin tuna in the Atlantic Ocean are the following:

1. In what geographic location would we see the highest recovery of tags? This is vital to making sure the experiment works (i.e., survival with the tag is ongoing).

2. Where and how should a tag be attached to a fish that has enormous

growth potential, i.e., externally or internally? On what year class should emphasis be placed?

3. How well will the geopositioning-light sensor technology that is irradiance based work in areas close to the equator? Efforts should be concentrated at latitudes where the technology is proven.

4. Develop programs for recovery of tags once placed on fish. Awareness of the value of the tag and its importance to scientists will require a large education effort.

5. Design a program that will yield maximum results.

APPENDIX
E

Evidence for Mixing Based on Parasites

Walters (1980)[1] examined the prevalence of two species of parasites from three locales (western Atlantic Ocean, Mediterranean Sea, and the Bay of Biscay). Based on the prevalence of the parasites, he argued that annual mixing of the two stocks was about 15%. He concluded that some school tuna in the Bay of Biscay "may" have been spawned in the Gulf of Mexico and recommended further study to support the postulate, however, to date, this has not been carried out.

Of all the techniques used to argue the case for mixing, this is by far the simplest. It requires a sample of 20 to 30 fish, ages 0 to 2, from each locale. The sample can consist of only the heads of the fish, and they can be preserved in 10% formalin.

Little is known about the life cycle of these parasites or how they are acquired. The life cycle is direct: eggs are shed into the water, and immatures are free-living. Immatures in the water column usually attach to the skin or gill cavity of the host in response to unknown chemical cues, and they migrate to the preferred site (the nasal cavity in the case of *Nasicola* sp.; the gill chamber in the case of *Elytrophora* sp.). Generally speaking, adult parasites of this type do not pass from one host to another. There is no information on the longevity of these parasites.

The analyses by others of Walters' data are not critical reviews. Murphy (1990) confuses prevalence and intensity and simply restates Walters' postulate

---

[1] All mention of Walters in this appendix refers to the work in Walters (1980).

that there is mixing of about 15% in both directions. Brunenmeister's (1980) analysis predates access to all of the Walters' data and so is incomplete. Bossert (1993) does not consider Walters' data. It is important then to restate Walters' view of his own data. He uses the terms "reasonable to postulate," "may have spawned," "a plausible explanation," "further study is indicated," "it appears that *Nasicola* sp. could be used," and "*Elytrophora* sp. appears to be acquired." Walters offers a possible explanation, but it is not more than that and always should be couched in those terms.

NASICOLA SP.

Most giant bluefin tuna in the western Atlantic Ocean are infested with *Nasicola* sp. (93% [Walters, 1980], 81% [Phipps, 1980]). None have been sampled from the eastern Atlantic Ocean.

The most definitive result in Walter's study is that mean parasite size increases with host size. Data of Phipps (1980) and Wheeler and Beverley-Burton (1987) confirm and extend this relation (Figure E-1). Walters suggested that the parasite may be very long lived because parasite size increases with host size and the host is long lived (20+ years). An alternative explanation, not uncommon in the parasitological literature, is that the parasites grow larger because they have more room to do so and that the giant fish repeatedly acquire the parasite each time they pass through water harboring immatures.

Table E-1 and Figure E-2A provide a summary of the data for *Nasicola* sp. The data of Phipps (i.e., worm size [1980]) suggest that the giant bluefin tuna used by Walters were probably small giant fish. Neither Walters nor Wheeler and Beverley-Burton (1987) give the host size of giant fish.

Walters used the data to suggest the following. The decrease in prevalence in the western Atlantic tuna (i.e., 0.86 at age 0 to 0.31 at age 3) was taken to indicate that the fish were being diluted by uninfested fish from the east; there is mixing. The presence of infested fish in the Bay of Biscay was taken to indicate that the uninfested fish from the Mediterranean Sea were being diluted by infested fish from the west; there is mixing.

The *Nasicola klawei* holotype (i.e., the specimens used to define the parasite species) was obtained from tuna other than bluefin (Yamaguti, 1963). Wheeler and Beverley-Burton (1987) proposed a new species (*N. hogansi*) for specimens obtained from western Atlantic bluefin tuna (Prince Edward Island, Canada). Their description is identical to that for a much larger sample described by Phipps (1980) also from western Atlantic giant bluefin tuna. It is likely the taxonomy is confused because workers have not sampled parasites from hosts of the same size. It is likely that the species from bluefin tuna described by Walters is not *N. klawei* but rather *N. hogansi*.

The high prevalence at age 4 was ignored by Walters; presumably it was taken to be anomalous. The slope of the line describing the decrease in preva-

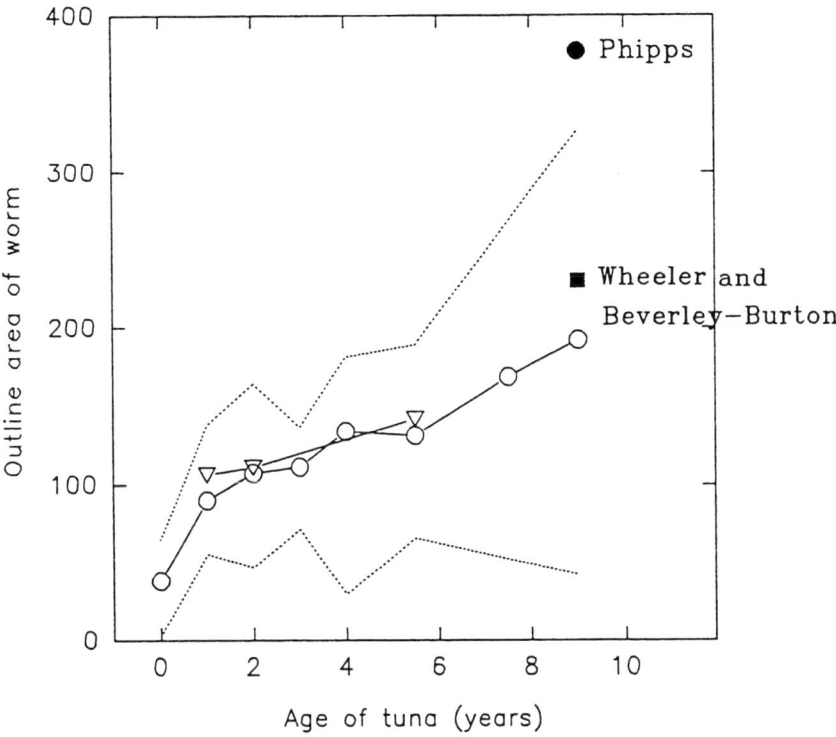

FIGURE E-1   The increase in size of *Nasicola* sp. with host size in Atlantic bluefin tuna. Age 9 on the x-axis actually refers to all fish 9 years or older (i.e., giant fish). Open circles are mean values for western bluefin tuna, and dashed lines indicate the range; open triangles are mean values for eastern bluefin tuna (Walters, 1980). Filled square (Wheeler and Beverley-Burton, 1987; tended line in Prince Edward Island) and filled circle (Phipps, 1980; fork length of 280 cm, traps in Nova Scotia) are mean values for western bluefin tuna.

lence from age 0 to age 8 is not significantly different from zero, thus weakening the dilution argument. The number of fish sampled from the east is very small. It is likely that the marked increase in prevalence between ages 8 and 9+ is related to a marked change in migration patterns. That is, at age 9 the giant fish return to warm water to spawn, but when they return to warm water they also acquire the parasite. An alternative and equally valid explanation of all of the data is that there are two regions where the parasite can be acquired, one in the west (in the region of the Florida Straits) and one in the east (near the Bay of Biscay). The prevalence data are some measure of the number of tuna passing through these regions at some time of the year, and the probability of a tuna

TABLE E-1 Prevalence of *Nasicola* sp. parasites in bluefin tuna related to age of the host and to locale. The fractions are the number of hosts infested divided by the number of hosts examined.

Age (years)	Western Atlantic Ocean	Mediterranean Sea	Bay of Biscay
0	19/22 = 0.86	0/9 = 0	—
1	19/25 = 0.76	0/2 = 0	2/14 = 0.14
2	38/62 = 0.61	—	2/24 = 0.08
3	8/26 = 0.31	0/1 = 0	—
4	19/24 = 0.79	0/3 = 0	0/3 = 0
5 & 6	3/17 = 0.18	0/1 = 0	1/12 = 0.08
7 & 8	9/25 = 0.36	—	—
9+	53/57 = 0.93	—	—
9+ Phipps	39/48 = 0.81	—	—

passing through the waters where they can acquire the parasite changes with their age.

ELYTROPHORA SP.

Figure E-2B and Table E-2 provide a summary of the data for the copepod.

Walters used these data to suggest that the western fish ages 2 to 5 infested with this parasite represent fish that moved from the east to the west.

The taxonomy on this parasite is also confused, but a revision has not been published (Phipps, 1980).

About 20% of young western tuna were infested with the parasite, and fish that were parasitized had about five parasites each. Walters argued that this is a nonrandom distribution (i.e., of 100 fish one would expect each to have one parasite rather than 20 fish each to have five parasites). He suggested that the infested and noninfested fish had different histories and that the infested fish in the west had their origin in the east. It is equally likely that the two groups had different histories but that the infested fish passed through waters harboring the parasite and others avoided these waters. It is possible that the waters harboring the parasite are in the west and not in the east as argued by Walters.

Except for age 0, sample sizes from the Mediterranean Sea are too small to allow one to draw any inferences.

Age 4 fish from the western Atlantic Ocean are not treated as an anomaly, whereas in the *Nasicola* sp. analysis they were. As with *Nasicola* sp., it is likely that the marked increase in prevalence in age 9+ is related to a marked change in migration patterns. That is, at age 9 the giant fish return to warm water to spawn but in doing so also pass through waters that allow them to acquire the parasite.

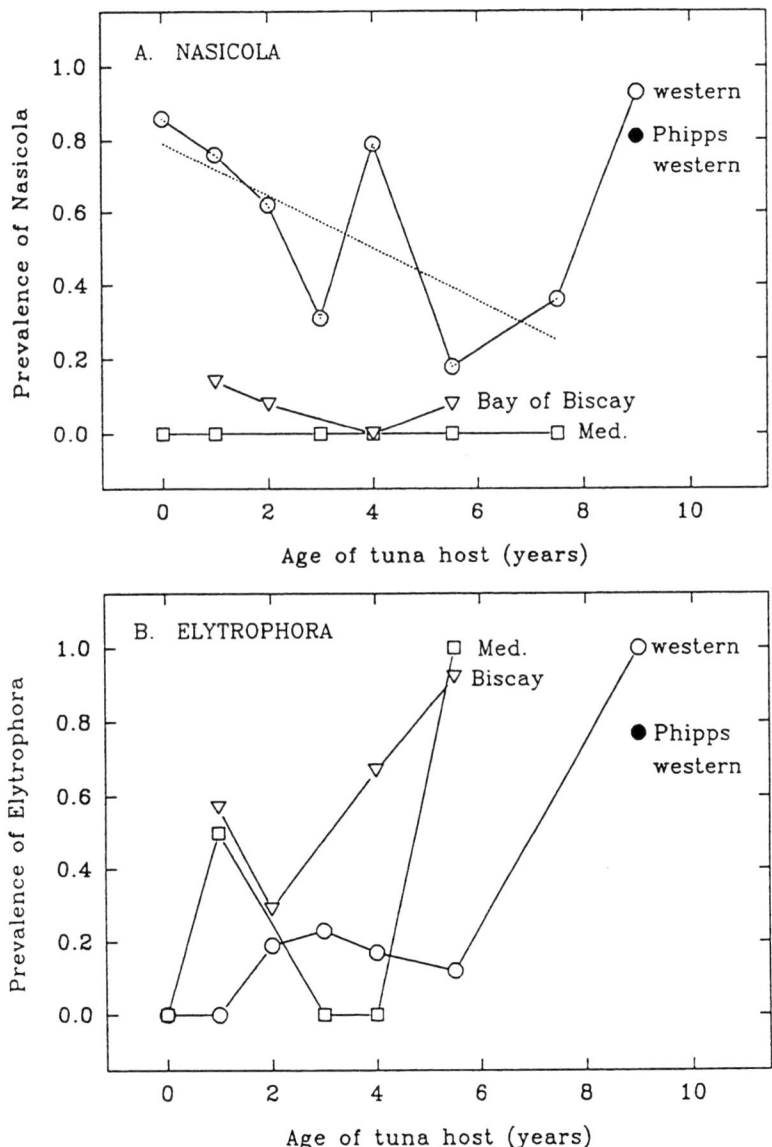

FIGURE E-2 Prevalence of parasites as a function of host age and locale of capture (A, *Nasicola* sp.; B, *Elytrophora* sp.). Age 9 on the x-axis actually refers to all fish 9 years or older (i.e., giant fish). Open circles are mean values for western tuna, and the dashed line is a linear regression for ages 0 to 8; open triangles are mean values for eastern bluefin tuna from the Bay of Biscay; open squares are mean values for eastern bluefin tuna from the Mediterranean Sea (Walters, 1980). Filled circles (Phipps, 1980; fork length of 280 cm, traps in Nova Scotia) are mean values for western bluefin tuna.

TABLE E-2 Prevalence of *Elytrophora* sp. parasites in bluefin tuna related to age of the host and to the locale of host capture. The fractions are the number of hosts infested divided by the number of hosts examined.

Age (years)	Western Atlantic Ocean	Mediterranean Sea	Bay of Biscay
0	0/22 = 0	0/9 = 0	—
1	0/25 = 0	1/2 = 0.5	8/14 = 0.57
2	12/62 = 0.19	—	7/24 = 0.29
3	6/26 = 0.23	0/1 = 0	—
4	4/24 = 0.17	0/3 = 0	2/3 = 0.67
5 & 6	2/17 = 0.12	1/1 = 1.0	11/12 = 0.92
7 & 8	—	—	—
9+	22/22 = 1.0	—	—
9+ Phipps	53/69 = 0.77	—	—

An alternative and equally valid explanation of all of the data is that there are two regions where the parasite can be acquired, one in the west (in the region of the Florida Straits) and one in the east (near the Bay of Biscay). The prevalence data are some measure of the number of tuna passing through these regions at some time of the year, and the probability of a tuna passing through the area where they can acquire the parasite changes with their age.

HOSTS WITH BOTH PARASITES

Walters reported that only three of 129 western hosts ages 2 to 6 were infested with both parasites. Table D-3 is a summary of his data. Walters incorrectly reported the sum of those infested with *Nasicola* sp. as 52; the sum is 68.

Walters argued that if the parasites were randomly distributed in a homoge-

TABLE E-3 Prevalence of *Nasicola* sp., *Elytrophora* sp., and both parasites in the same host for western Atlantic bluefin tuna ages 2 to 6.

Age (years)	*Nasicola* sp.	*Elytrophora* sp.	Both parasites
2	38/62	12/62	Not reported
3	8/26	6/26	Not reported
4	19/24	4/24	Not reported
5 & 6	3/17	2/17	Not reported
Sum	68/129 = 0.53	24/129 = 0.19	3/129 = 0.02

neous host stock, one would expect 68 × 24/129 = 13 fish to harbor both parasites, whereas he observed only three. Walters draws the inference that the host stock is not homogeneous. It is equally likely that the parasites simply are not distributed randomly.

SUMMARY

One satisfactory explanation for the data is that there is mixing of young tuna: about 15% of tuna moving from east to west and a similar number moving in the other direction.

An equally valid explanation is that western bluefin tuna acquire both parasites in the relatively warm waters in the west, that eastern bluefin tuna can acquire both parasites in the relatively warm waters of the east, and that the water temperature that permits acquisition is different for the two parasites. The Walters study does not provide evidence to support mixing.

REFERENCES

Bossert,[2] 1993. Review of the ICCAT bluefin tuna working group report, draft of 10/2/93, and supporting documents.

Brunenmeister, S. 1980. A summary and discussion of technical information pertaining to the geographical discreteness of Atlantic bluefin tuna resources. ICCAT Coll. Vol. Sci. Pap. IX:506-527 (SCRS/79/95).

Murphy, G.I. 1990. A review of the evidence of stock structure in Atlantic bluefin tuna with an alternate stock hypothesis. Draft.

Phipps, M. 1980. Preliminary studies of population structure and mortality of bluefin tuna (*Thunnus thynnus*) in Nova Scotia, Canada. MSc thesis, University of Guelph.

Walters, V. 1980. Ectoparasites of eastern and western Atlantic bluefin tuna. ICCAT Coll. Vol. Sci. Pap. IX(2): 491-498 (SCRS/79/79).

Wheeler, T.A., and M. Beverley-Burton. 1987. *Nasicola hogansi* n.sp. (Monogenea: Capsalidae) from bluefin tuna, *Thunnus thynnus* (Osteichthyes: Scombridae), in the northwest Atlantic. Can. J. Zool. 65:1947-1950.

Yamaguti, S. 1963. Monogenetic Trematodes of Hawaiian Fishes. University of Hawaii Press, Honolulu. 287pp.

---

[2]William Bossert, Harvard University, Cambridge, MA 02138.

APPENDIX
F

Microconstituent Analysis

The committee received three documents to review (Calaprice, 1983, 1984, 1985); none has been published in the primary literature. Other material that we received did not contain any critical reviews of the Calaprice study. The reviews that we received tended to restate the generalizations in Calaprice (1985).

The major difficulty with the Calaprice study is that different statistical techniques led to different mixing values, and that repeated analysis on the same material led to different mixing values.

For example, in a pilot study, Calaprice used 36 age 2+ fish from the Bay of Biscay and 39 age 2+ fish from the coast of Virginia and statistically analyzed various sections of the x-ray spectra separately with a stepwise discriminant function analysis. Out of 512 bins (energy levels), the stepwise discriminant function analysis found eight variables that gave 100% discrimination between the eastern and western Atlantic samples. Calaprice concluded that "it was possible to derive equations that could be used to classify individuals as to area of origin." These early results, however, imply that there was no mixing among age 2+ fish from the two areas, which was contradicted by later analyses of samples from the same areas that apparently showed that some mixing was occurring.

One case where Calaprice deals with uncertainties in his estimates is in the analysis of his pilot study mixing of 39 age 2+ bluefin tuna from the western Atlantic Ocean and 36 age 2+ bluefin tuna from the eastern Atlantic Ocean (Calaprice, 1983, Table 3). Mixing from west to east is given as 14.3%, but the 95% confidence levels are 3 to 37%. Mixing from east to west is given as 23.8%, but the 95% confidence levels are 11 to 53%.

TABLE F-1 Estimates of mixing between the two stocks based on discriminant function analysis using jackknife probabilities of group membership of adult bluefin tuna sampled from a variety of locations and at different times.

Source of tuna	Estimated % mixing	
	Analyzed 1983	Analyzed 1984
Western Atlantic Ocean		
Massachusetts (1980)	12.7	10.8
Massachusetts (1982)	—	11.9
Gulf of Mexico (1983)	—	4.2
Eastern Atlantic Ocean		
Gibraltar (1982)	11.3	90.0
Gibraltar (1983)	—	4.3
Tyrrhenian Sea (1982)	12.9	60.0
Tyrrhenian Sea (1983)	—	5.2
Ionian Sea (1982)	14.3	100.0

A summary of some other problems is presented in Table F-1 (adapted from Calaprice, 1983, Table 2). Estimated mixing varies from 4 to 100%, depending on when it was analyzed and on the source of the material.

Although there are new pattern recognition techniques that could be applied to the data collected by Calaprice, the committee does not recommend such analysis. A better approach would be to apply new analytical techniques to new samples.

REFERENCES

Calaprice, J.R. 1983. X-ray fluorescence of stock variation in bluefin tuna, status report. pp. 1-42.
Calaprice, J.R. 1984. X-ray fluorescence study of stock variation in bluefin tuna, Third quarterly report. pp. 1-25.
Calaprice, J.R. 1985. Chemical variability and stock variation in northern Atlantic bluefin tuna. ICCAT SCRS/85/36. pp. 222-252.